sustainability

NATIONAL GEOGRAPHIC LEARNING | **DELMAR** CENGAGE Learning

Australia • Brazil • Japan • Korea • Mexico • Singapore • Spain • United Kingdom • United States

Sustainability

Vice President, Career, Education, and Training Editorial: David Garza

Director of Learning Solutions: Sandy Clark

Managing Editor: Larry Main

Associate Acquisitions Editor: Katie Hall

Editorial Assistant: Kaitlin Murphy

Product Manager: Anne Prucha

Instructional Designer: Nancy Pettit

Media Editor: Debbie Bordeaux

Associate Marketing Manager: Jillian Borden

Production Director: Wendy Troeger

Production Manager: Mark Bernard

Senior Content Project Manager: Glenn Castle

Director of Design: Bruce Bond

Manufacturing Planner: Beverly Breslin

Production and Composition: Integra

Text and Cover Designer: Bruce Bond

Cover Image: Michael Melford

National Geographic Image Collection

For product information and technology assistance, contact us at **Cengage Learning Customer & Sales Support, 1-800-354-9706.**

For permission to use material from this text or product, submit all requests online at **www.cengage.com/permissions.** Further permissions questions can be e-mailed to **permissionrequest@cengage.com.**

Library of Congress Control Number: 2012944691

ISBN-13: 978-1-285-06061-3
ISBN-10: 1-285-06061-X

Delmar
5 Maxwell Drive,
Clifton Park, NY 12065-2919
USA

Cengage Learning is a leading provider of customized learning solutions with office locations around the globe, including Singapore, the United Kingdom, Australia, Mexico, Brazil, and Japan. Locate your local office at **www.cengage.com/global**.

Cengage Learning products are represented in Canada by Nelson Education, Ltd.

To learn more about Delmar, visit **www.cengage.com/delmar**
Purchase any of our products at your local college store or at our preferred online store **www.CengageBrain.com.**

Printed in the United States of America
1 2 3 4 5 6 7 16 15 14 13 12

Table *of* Contents

About the Series

Cengage Learning and National Geographic Learning are proud to present the *National Geographic Learning Reader Series*. This ground breaking series is brought to you through an exclusive partnership with the National Geographic Society, an organization that represents a tradition of amazing stories, exceptional research, first-hand accounts of exploration, rich content, and authentic materials.

The series brings learning to life by featuring compelling images, media, and text from National Geographic. Through this engaging content, students develop a clearer understanding of the world around them. Published in a variety of subject areas, the *National Geographic Learning Reader Series* connects key topics in each discipline to authentic examples and can be used in conjunction with most standard texts or online materials available for your courses.

How the reader works

Each article is focused on one topic relevant to the discipline. The introduction provides context to orient students and focus questions that suggest ideas to think about while reading the selection. Rich photography, compelling images, and pertinent maps are amply used to further enhance understanding of the selections. The article culminating section includes discussion questions to stimulate both in-class discussion and out-of-class work.

A premium eBook will accompany each reader and will provide access to the text online with a media library that may include images, videos and other premium content specific to each individual discipline.

National Geographic Learning Readers are currently available in a variety of course areas, including Archeology, Architecture & Construction, Biological Anthropology, Biology, Earth Science, English Composition, Environmental Science, Geography, Geology, Meteorology, Oceanography and Travel and Tourism.

Few organizations present this world, its people, places, and precious resources in a more compelling way than National Geographic. Through this reader series we honor the mission and tradition of National Geographic Society: to inspire people to care about the planet.

I t is an exciting time for the world as we begin to adopt natural, sustainable practices. Countries around the planet have risen, and continue to adapt, to meet the challenge of embracing a sustainable lifestyle. From solar panels to wind turbines, a new vocabulary is replacing the way we speak about energy. The articles in this reader are a testament to the work that has been done thus far in the field of sustainability and show that while challenges continue to arise, we are making headway and beginning to see our clean energy efforts come to fruition. The powerful images displayed in this Reader—green roofs, energy-efficient buildings, and hundreds of solar panels—are just the start. As Michael Parfit (contributing author) says, "The excitement of energy freedom seems contagious." Throughout these articles, readers are asked to explore the ways each individual can make a difference in the decisions that they make surrounding the usage of energy and resources. The articles and images presented herein remind us that we are not here to consume; instead, we can leave a legacy of sustainable choices.

To begin, readers investigate China's paradoxical role in the green movement in "Can China Go Green?" While the country is the planet's leader in renewable energy technology, it also burns more coal than any other country. Next, readers travel to Iceland to explore the decision between creating clean energy sources and keeping the landscape pristine in "Iceland's Power Struggle." Solar energy issues are then discussed with the following two articles, "Plugging into the Sun" and "Can Solar Save Us?" "The 21st Century Grid" examines the dated power infrastructure that links us all together. To truly validate our green energy efforts, the grid must be connected to these renewable energy sources. Next, the conversation moves from clean energy to clean water in "The Big Idea: Get the Salt Out" and "The Gulf of Oil: The Deep Dilemma." Readers examine the future in "Next: Space Elevator" which discusses ways that we could make life in space easier with technology advancement as the world expands beyond the traditional boarders. Finally, the last two articles present sustainable architecture through the articles "Environment: London's Green Giant" and "Up on the Roof."

This National Geographic reader brings together a diverse group of investigations into the natural resources and technological developments that will lessen our carbon footprint on the globe. Each article challenges the student to think beyond the content in the article and to consider how these advances impact them closer to home. An Anticipation Guide introduces each article which then concludes with a Career Investigation, Team Building Activity, and a Writing Assignment. Instructors can choose additional activities which are available on the Instructor Companion website. The Career Investigation highlights careers relevant to the individual articles and allows readers the chance to further explore the educational requirements, essential skills needed, job descriptions, and pay scale of a variety of occupations. The Team Building Activity asks four or five people to form a team and research a topic further. The Writing Assignment asks for further thought about the article and the impact that it has locally. These up-to-date and relevant *National Geographic* articles help students gain a perspective on how the choices they make daily can impact their local community, environment, and even the world.

© Michael Melford/National Geographic Stock

Purpose: To identify what you already know about living "green," to direct and personalize your reading, and to provide a record of what new information you learned.

———————◆———————

Before you read "Can China Go Green?" examine each statement below and indicate whether you agree or disagree. Be prepared to discuss your reactions to the statements in groups.

- China's economy has grown steadily (more than 8 percent) for the past 10 years.
- China currently leads the world in greenhouse gas emissions.
- China is leading the world in utilizing solar energy in an urban setting.
- China is the leading producer of wind energy in the world.
- China has the lowest greenhouse emissions (air pollution) in the world.
- As China's economy continues to grow along with the production of wind and solar energy, so does the pollution problem.
- Since China's economy has increased on average 8–10 percent for the past ten years, everyone in China is happy with the progress.
- China's expanding economy and steps to go green include the expansion of subway lines and high speed trains (in more than 25 cities), thereby reducing the amount of pollution in the skies.
- Many of China's solar panels operate at 50 percent capacity.
- China is leading the world in becoming a green country.
- China's use of coal to produce electricity is expected to continue growing through at least 2030.

CAN CHINA GO GREEN?

By Bill McKibben

Henan, People's Republic of China.

NO OTHER COUNTRY
IS INVESTING SO HEAVILY
IN CLEAN ENERGY.
BUT NO OTHER COUNTRY BURNS AS MUCH COAL
TO FUEL ITS ECONOMY.

Rizhao, in Shandong Province, is one of the hundreds of Chinese cities gearing up to really grow. The road into town is eight lanes wide, even though at the moment there's not much traffic. But the port, where great loads of iron ore arrive, is bustling, and Beijing has designated the shipping terminal as the "Eastern bridgehead of the new Euro-Asia continental bridge." A big sign exhorts the residents to "build a civilized city and be a civilized citizen."

In other words, Rizhao is the kind of place that has scientists around the world deeply worried—China's rapid expansion and new-found wealth are pushing carbon emissions ever higher. It's the kind of growth that helped China surge past the United States in the past decade to become the world's largest source of global warming gases.

And yet, after lunch at the Guangdian Hotel, the city's chief engineer, Yu Haibo, led me to the roof of the restaurant for another view. First we clambered over the hotel's solar-thermal system, an array of vacuum tubes that takes the sun's energy and turns it into all the hot water the kitchen and 102 rooms

Solar is in at least 95 percent of all the buildings.

can possibly use. Then, from the edge of the roof, we took in a view of the spreading skyline. On top of every single building for blocks around a similar solar array sprouted. Solar is in at least 95 percent of all the buildings, Yu said proudly. "Some people say 99 percent, but I'm shy to say that."

Whatever the percentage, it's impressive—outside Honolulu, no city in the U.S. breaks single digits or even comes close. And Rizhao's solar water heaters are not an aberration. China now leads the planet in the installation of renewable energy technology—its turbines catch the most wind, and its factories produce the most solar cells.

We once thought of China as the "yellow peril" and then the "red menace." Now the colors are black and green. An epic race is on, and if you knew how the race would come out—if you knew whether or how fast China could wean itself off coal and tap the sun and wind—then you'd have the single most important data point of our century. The outcome

Adapted from "Can China Go Green?" by Bill McKibben: National Geographic Magazine, June 2011.

of that race will determine how bad global warming is going to get. And right now the answer is still up in the air.

Literally up in the air. Visitors to China are instantly struck, of course, by the pollution shrouding every major city. Slowly those skies are clearing a little, at least in places like Beijing and Shanghai, as heavy industry is modernized or moved out of town. And the government has shut down many of the smallest and filthiest coal-fired power plants. Indeed, the country now leads the world in building what engineers call supercritical power stations, which produce far less smog than many of the hulking units still online in the U.S. Presumably China will get steadily cleaner as it gets richer—that's been the story elsewhere.

But—and it's a crucial but—you can clean the air without really cleaning the air. The most efficient coal-fired power plants may not pour as much particulate matter, sulfur dioxide, and nitrogen oxides into the atmosphere, but they still produce enormous quantities of carbon dioxide. Invisible, odorless, generally harmless to humans—and the very thing that's warming the planet. The richer China gets, the more it produces, because most of the things that go with wealth come with a gas tank or a plug. Any Chinese city is ringed with appliance stores; where once they offered electric fans, they now carry vibrating massage chairs.

"People are moving into newly renovated apartments, so they want a pretty, new fridge," a clerk told me. "People had a two-door one, and now they want a three-door." The average Shanghainese household already has 1.9 air conditioners, not to mention 1.2 computers. Beijing registers 20,000 new cars a month. As Gong Huiming, a transportation program officer at the nonprofit Energy Foundation in Beijing, put it: "Everyone wants to get the freedom and the faster speed and the comfort of a car."

That Chinese consumer revolution has barely begun. As of 2007, China had 22 cars for every 1,000 people, compared with 451 in the U.S. Once you leave the major cities, highways are often deserted and side roads are still filled with animals pulling carts. "Mostly, China's concentrated on industrial development so far," said Deborah Seligsohn, who works in Beijing for the Washington, D.C.-based World Resources Institute. Those steel mills and cement plants have produced great clouds of carbon, and the government is working to make them more efficient. As the country's industrial base matures, their growth will slow. Consumers, on the other hand, show every sign of speeding up, and certainly no Westerner is in a position to scold.

Bill Valentino, a sustainability executive with the pharmaceutical giant Bayer who has long been based in Beijing, recently taught a high school class at one of the international schools. He had his students calculate their average carbon footprint, and they found that if everyone on the planet lived as they did, it would take two to four Earths' worth of raw materials to meet their needs. So they were already living unsustainable lives. Valentino—an expat American who flies often—then did the same exercise and found that if the whole world adopted his lifestyle, we'd require more than five planet Earths.

China has made a low-carbon economy a priority, but no one is under any illusion about the country's chief aim. By most estimates, China's economy needs to grow at least 8 percent a year to ensure social stability and continued communist rule. If growth flags, Chinese may well turn rebellious; there are estimates of as many as 100,000 demonstrations and strikes already each year. Many of them are to protest land takeovers, bad working conditions, and low wages, so the government's main hope is to keep producing enough good jobs to absorb a population still pouring out of the poor provinces with high hopes for urban prosperity.

Increasingly, though, Chinese anger is directed at the environmental degradation that has come with that growth. On one trip I drove through a village north of Beijing where signs strung across the road decried a new gold mine for destroying streams. A few miles later I came to the mine itself, where earlier that day peasants had torn up the parking lot, broken

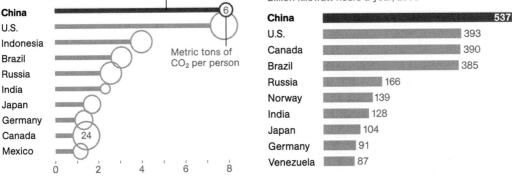

Greenhouse gas emissions
Total emissions in gigatons of CO_2, 2007

China	6
U.S.	
Indonesia	
Brazil	
Russia	
India	
Japan	
Germany	
Canada	24
Mexico	

Metric tons of CO_2 per person

0   2   4   6   8

Electricity production from renewable sources
Billion kilowatt-hours a year, 2008

China	537
U.S.	393
Canada	390
Brazil	385
Russia	166
Norway	139
India	128
Japan	104
Germany	91
Venezuela	87

Change (and CO_2) in the Air
Burning more than three billion tons of coal a year—more than the U.S., Europe, and India combined—China tops the world in emissions of CO_2 and other atmosphere-warming gases. To slow emissions without impeding its supercharged economic growth, the nation has also become a leader in clean energy, generating nearly 20 percent of its electricity from renewable sources, mostly hydro and wind.

LAWSON PARKER AND KAITLIN M. YARNALL, NGM STAFF

SOURCES: MCKINSEY & COMPANY; POPULATION REFERENCE BUREAU; U.S. ENERGY INFORMATION ADMINISTRATION

the windows, and scrawled graffiti across the walls. A Chinese government-sponsored report estimates that environmental abuse reduced the country's GDP growth by nearly a quarter in 2008. The official figures may say the economy is growing roughly 10 percent each year, but dealing with the bad air and water and lost farmland that come with that growth pares the figure to 7.5 percent. In 2005 Pan Yue, vice minister of environmental protection, said the country's economic "miracle will end soon, because the environment can no longer keep pace." But his efforts to incorporate a "green GDP" number into official statistics ran into opposition from Beijing.

"Basically," said one Beijing-based official who refused to be identified (itself a reminder of how sensitive these topics are), "China seeks every drop of fuel—every kilowatt and every kilojoule it can get a hold of—for growth." So the question becomes, What will that growth look like?

One thing it already looks like is: big and empty. Ordos, in Inner Mongolia, may be the fastest growing city in China; even by Chinese standards it has an endless number of construction cranes building an endless number of apartment blocks. The city's great central plaza looks as large as Tiananmen Square in Beijing, and towering statues of local-boy-made-good Genghis Khan rise from the concrete plain, dwarfing the few scattered tourists who have made the trek here. There's a huge new theater, a modernist museum, and a remarkable library built to look like leaning books. Coal built this Dubai-on-the-steppe. The area boasts one-sixth of the nation's total reserves, and as a result, the city's per capita income had risen to $20,000 by 2009. (The local government has set a goal of $25,000 by 2012.) It's the kind of place that needs some environmentalists.

And indeed it has at least one. In the neighboring city of Baotou, a steelmaking center whose mines also supply half the planet's rare earth minerals, I found Ding Yaoxian ensconced in the headquarters of the nonprofit Baotou City Environment Federation, on the second floor of a day center for retired party cadres, who were playing badminton on the mezzanine. Director Ding is one of the most cheerful and engaging Chinese I've ever met; he's needed every bit of charisma to

Solar panels on top of the Yingli Solar Company factory.

build his association into a real force, numbering by his account a million area citizens. Issued little green identity cards, they serve as a kind of volunteer police force. "If people from the association see someone spilling trash, they go and sit on their doorstep," Ding said. "The government can't have eyes everywhere. A voluntary organization can put more pressure on. It can shame."

But the campaigns the group focuses on most of the time make clear how nascent environmental concern in China still is. They've handed out a million reusable shopping bags—but also hundreds of thousands of small folding paper cups, so that people will stop spitting on the street. One minor victory: When showing those hundreds of thousands of new condo units, real estate agents used to hand customers plastic

booties to go over their dirty shoes; now they supply washable cloth socks. The association has tried to introduce the concept of garage sales, in a country where secondhand goods carry a stigma. And members have launched a big effort to teach Inner Mongolians to smile. "In the West people are happy and smiling, and that makes people feel positive," Ding said. His deputy, Feng Jingdong, added, "We tell them, use your personality to get people to enjoy themselves instead of using resources." The three of us were eating a delicious lunch at a nearby restaurant (lamb is the staple here), and when we were finished, Ding made sure to ask for a doggie bag. "That's one of our campaigns," he said. "Before, it felt like you lost face if you did that."

There was one truly significant sign of greening long under way in *(Continued on page 11)*

China's power plants, 2010

Renewable-energy plants are numerous,
but most power is still generated by coal.

● One power plant

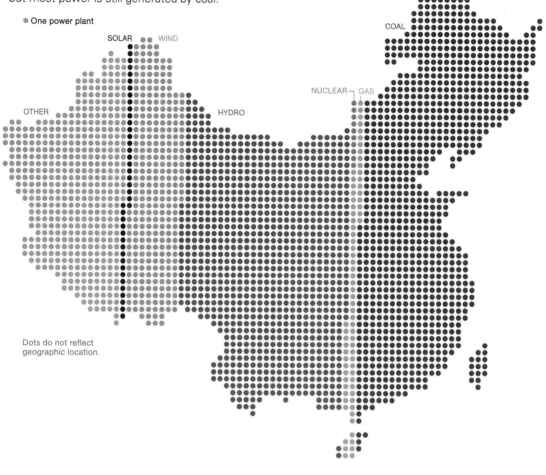

SOLAR ●● WIND

COAL

NUCLEAR ― GAS

OTHER

HYDRO

Dots do not reflect
geographic location.

China's power generation, terawatt-hours

						Total
2005						**2,503**
2030						**9,256**

Other ⌐ ⌐ Hydro   Nuclear   Gas   Coal
⌐ Wind
Solar

A Coal-fueled Country

Despite aggressive growth in alternative energy, China will
have to burn even more coal to power its booming economy.
A forecast for 2030 (above) indicates that if China meets
current policy commitments, coal will continue to supply
70 percent of its energy, double the amount used now. To
emit no more greenhouse gases in 2030 than today, China
would have to rely on green energy for two-thirds of its
power generation.

NOTE: ONE TERAWATT-HOUR = 1 BILLION KILOWATT-
HOURS. ONE KILOWATT-HOUR WILL POWER A
100-WATT LIGHTBULB FOR TEN HOURS.

LAWSON PARKER AND KAITLIN M. YARNALL,
NGM STAFF

SOURCES: PLATTS/UDI WORLD ELECTRIC POWER
PLANTS DATABASE; MCKINSEY & COMPANY

(Continued from page 9) the region: a massive tree-planting campaign designed to hold the fragile soil in place. Flatbed trucks packed with seedlings were the second most common sight along area roads (outnumbered ten to one, it seemed, by trucks carrying coal from the mines). Ding estimated that he'd planted 100,000 trees with his own hands. "It used to be very dusty here, with lots of sandstorms," he said. "But we had 312 blue sky days last year, and every year there are more."

In search of further reassurance that China's booming growth held real seeds of environmental possibility, I drove 170 miles south of Beijing to the (redundancy alert) booming city of Dezhou. Approaching along National Highway 104, I got a sudden glimpse of one of the world's most remarkable buildings, the Sun-Moon Mansion. It looks like a convention center surrounded by the rings of Saturn, great tracks of solar panels providing all its hot water and electricity. Behind the hotel, a sister building serves as the headquarters of Himin Solar Corporation, which claims to have installed more renewable energy than any other company on Earth. (Chinese enterprises are sometimes the beneficiaries of largesse from Beijing, such as low-interest loans that may never need to be repaid in full.)

Himin's main products are those humble solar-thermal tubes that covered the rooftops in Rizhao. And as it turns out, they cover a lot of other real estate. Huang Ming, who founded the company, estimates that it's erected more than 160 million square feet of solar water heaters. "That means 60 million families, maybe 250 million people altogether—almost the population of the United States," he said. Huang, an ebullient fellow in faded black Dockers who used to be a petroleum engineer, sells some of the best solar-thermal systems in China, but even he admits that it's fairly simple technology. He says that the key to his company's success

China's green effort is being overwhelmed by the sheer scale of the coal-fueled growth. It's a dark picture.

has been opening people's minds, which it's done with revival-style marketing campaigns that storm one city at a time. "We do road showing, lecturing, PowerPointing," he said. And now they're harnessing the power of sightseeing too: The Sun-Moon Mansion is merely the anchor of a vast solar city that will soon include a solar "4-D" cinema, a solar video-game hall, a huge solar-powered Ferris wheel, and solar-powered boats to rent from a solar marina.

The company showroom, Feel It Hall, captures a few contradictions. The solar panels heat water for hot tubs and have giant flat-screen TVs above each. But that's the only way to sell the idea of renewable energy, Huang insisted, as he described the gigantic apartment towers he's building on the edge of town, with racks of solar panels that curve like the back of a dragon. "At night that's what you see—a floating dragon," he said. "So many developers come to our Solar Valley to copy from us, to learn from us. That's just what I wanted."

He's especially happy that some of those visitors come from abroad. Dezhou hosted the International Solar Cities World Congress in 2010, and he's set up an international-experts mansion for visiting dignitaries. "If all the people of the U.S.A. enjoyed solar hot water, Obama would win five Nobel Prizes!" he said. But it's going to take a while for America to catch up. Most of the U.S.'s minuscule capacity is used to heat swimming pools. Jimmy Carter had solar water heaters installed on the roof of the White House in 1979, but they came down in the Ronald Reagan years; new ones are due to go up this year.

It's not the only instance of the Chinese taking an American lead and running with it. Suntech has emerged as one of the top two leading makers of solar photovoltaic panels in the world. New employees are added weekly, and on their first day on the job they watch Al Gore in *An Inconvenient Truth*. The young tour guide showing me around the company's headquarters in Wuxi, near Shanghai, paused by the photos of solar panels at base camp on Mount Everest and the portrait of her boss, Shi Zhengrong, named by *Time* as one of its "heroes of the environment." "It's not only a job," she said, a tear welling in her eye. "I have…mission!"

Of course, that tear might have come in part from the air. Wuxi was among the dirtiest cities I'd ever visited: The 100-degree-Fahrenheit air was almost impossible to breathe. The solar array that forms the front of the Suntech headquarters slanted up to catch the sun's rays. Because of the foul air, it operated at only about 50 percent of its potential output.

In the end, anecdote can take you only so far. Even data are often suspect in China, where local officials have a strong incentive to send rosy pictures off to Beijing. But here's what we know: China is growing at a rate no big country has ever grown at before, and that growth is opening real opportunities for environmental progress. Because it's putting up so many new buildings and power plants, the country can incorporate the latest technology more easily than countries with more mature economies. It's not just solar panels and wind turbines. For instance, some 25 cities are now putting in or expanding subway lines, and high-speed rail tracks are spreading in every direction. All that growth takes lots of steel and cement and hence pours carbon into the air—but in time it should drive down emissions.

That green effort, though, is being overwhelmed by the sheer scale of the coal-fueled growth. So for the time being, China's carbon emissions will continue to soar. I talked with dozens of energy experts, and not one of them predicted emissions would peak before 2030. Is there anything that could move that 2030 date significantly forward? I asked one expert in charge of a clean-energy program. "Everyone's looking, and no one is seeing anything," he said.

W e took in a view of the skyline. On top of every building for blocks around a solar array sprouted.

Even reaching a 2030 peak may depend in part on the rapid adoption of technology for taking carbon dioxide out of the exhaust from coal-fired power plants and parking it underground in played-out mines and wells. No one knows yet if this can be done on the scale required. When I asked one scientist charged with developing such technology to guess, he said that by 2030 China might be sequestering two percent of the carbon dioxide its power plants produce.

Which means, given what scientists now predict about the timing of climate change, the greening of China will probably come too late to prevent more dramatic warming, and with it the melting of Himalayan glaciers, the rise of the seas, and the other horrors Chinese climatologists have long feared.

It's a dark picture. Altering it in any real way will require change beyond China—most important, some kind of international agreement that transforms the economics of carbon. At the moment China is taking green strides that make sense for its economy. "Why would they want to waste energy?" Deborah Seligsohn of the World Resources Institute asked, adding that "if the U.S. changed the game in a fundamental way—if it really committed to dramatic reductions—then China would look beyond its domestic interests and perhaps go much further." Perhaps it would embrace more expensive and speedier change. In the meantime China's growth will blast onward, a roaring fire that throws off green sparks but burns with ominous heat.

"To change people's minds is a very big task," Huang Ming said as we sat in the Sun-Moon Mansion. "We need time, we need to be patient. But the situation will not give us time." A floor below, he's built a museum for busts and paintings of his favorite world figures: Voltaire, Brutus, Molière, Michelangelo, Gandhi, Pericles, Sartre. If he—or anyone else—can somehow help green beat black in this epic Chinese race, he'll deserve a hallowed place near the front of that pantheon.

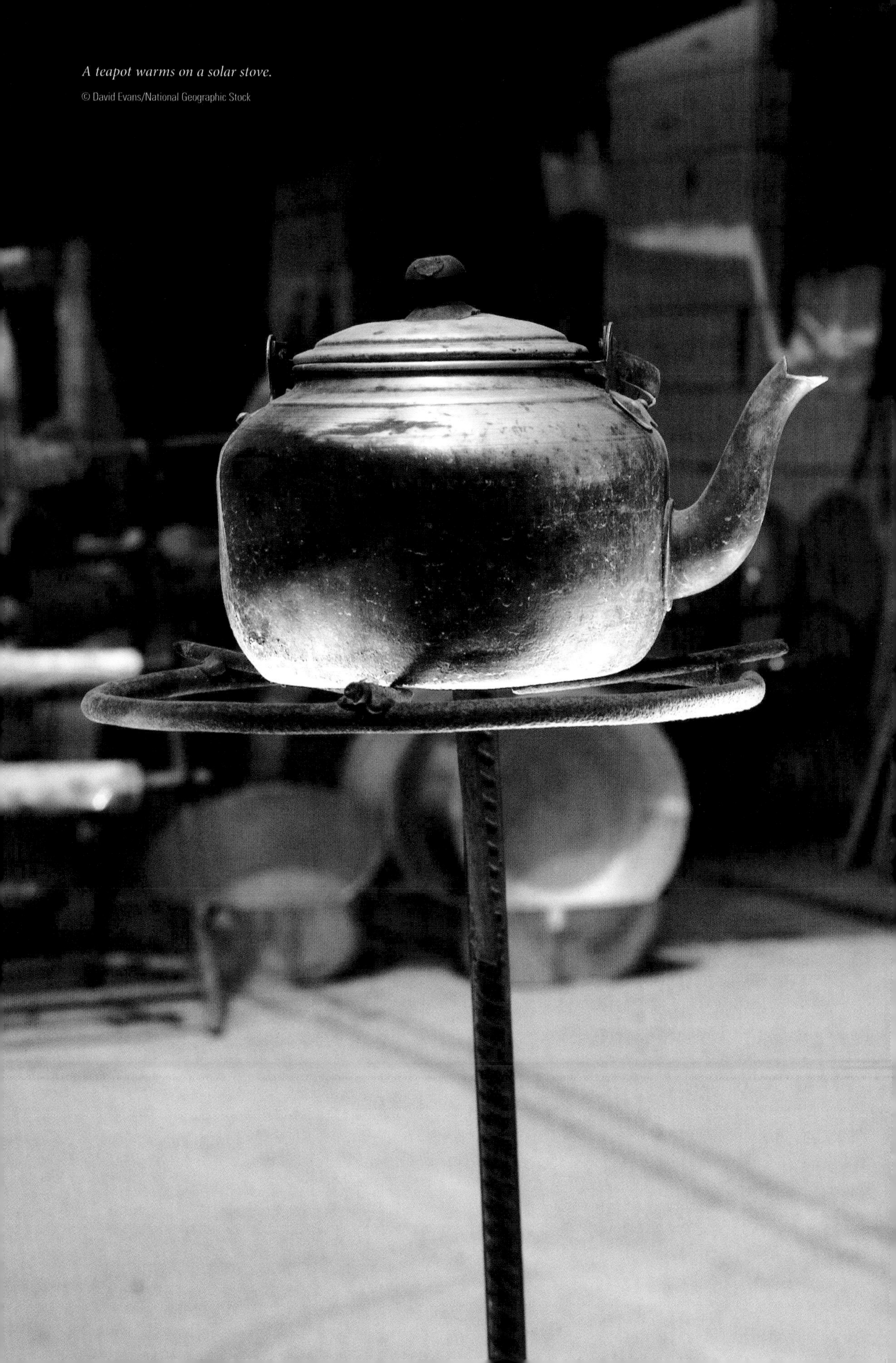

A teapot warms on a solar stove.

Career Investigation

China's economy has been growing between 8 and 10 percent annually for the past ten years. With that growth, China has been trying to improve its sustainability and become a "green" country. The increasing economy also means increasing production and, unfortunately, increasing greenhouse emissions.

Listed here are several possible careers to investigate. You may also be able to find a career path not listed. Choose three possible career paths and investigate what you would need to do to be ready to fill one of those positions.

Among the items to look for:

- Education—Is a college degree required? What would you major in? Should you have a minor?

- Working conditions—What is the day to day job like? Will you have to be out in the field all day, or is it a desk job? Is the work strenuous? Are you working on a drilling rig at sea, away from your family for 3–6 months at a time? Is the job "hazardous"?

- Pay scale—What is the average "starting pay" for the position you are seeking? Don't be fooled by looking at "average pay" which may include those who have been working for 20+ years.

- Is the job located in the states, or is there a possibility to travel to other parts of the world?

- Are there any other special requirements?

Architect
Architectural Technician or Technologist
Biochemist
Biologist
Biomedical Scientist
Biotechnologist
Botanist
Chemical Engineer
Chemical Engineering Technician
Chemist
Civil Engineer
Civil Engineering Technician
Clerk of Works
Commercial Energy Assessor
Ecologist
Electrician
Electricity Distribution Worker
Electronics Engineer
Engineering Construction Craftworker
Engineering Construction Technician
Gas Network Engineer
General Practice Surveyor
Geoscientist
Geotechnician
Landscape Manager
Landscape Scientist
Landscaper
Microbiologist
Minerals Surveyor
Nuclear Engineer
Oceanographer
Operational Researcher
Physicist
Refrigeration and Air Conditioning Engineer
Research Scientist
Structural Engineer
Thermal Insulation Engineer
Town Planner
Town Planning Support Staff
Water Network Operative

Team Building Activity

You will be assigned to teams of four or five students and host a panel that is composed of experts from around the world in the field of green technologies. The panel will examine the progress China is making in becoming green. The team will use PowerPoint, posters, videos, and other supporting resources to solidify their points. Each team member should cover a specific green technology (some examples are listed below) that China is currently pursuing. After the presentation, the audience (representing members of the media and/or experts from the field) will be allowed to question the panel.

- Energy Conservation
- Electric Vehicles
- Wave Energy
- Hydroelectricity
- Wind Power
- Hydrogen Fuel Cell
- Ocean Thermal Energy Conversion
- Solar Power

Writing Assignment

After reading the article "Can China Go Green?" and conducting your own research, write a brief paper discussing the pros and cons of China's steps toward becoming a "green" country.

In addition to your own ideas and thoughts, please discuss:

- What are other countries doing to move toward sustainability (going green)?
- Are all of China's power plants "dirty" coal-fired plants?

Listed here are a few links to help you in getting started.

Links:

Reporter's Notebook: China Goes Green
http://abcnews.go.com/International/China/china-green-focusing-energy-technology-transportation/story?id=12160354#.T6p1ecVlPis

China Goes Green in 2011
http://www.isuppli.com/china-electronics-supply-chain/marketwatch/pages/china-goes-green-in-2011.aspx

The Ozone Layer Fact Sheet
http://www.theozonehole.com/fact.htm

Ozone Is Found to Curb Plants' Powers
http://articles.latimes.com/2007/jul/26/science/sci-ozone26

China Goes Green with Rare Earth Plea
http://www.theregister.co.uk/2012/04/26/china_green_wto_rare_earth/

As China Goes Green What Is Canada Waiting for?
http://www.huffingtonpost.ca/john-brian-shannon/china-green_b_1150979.html

China Goes Green Initiating Major Climate Change Project
http://usgreentechnology.com/stories/china-goes-green-initiating-major-climate-change-project/

Greenhouse Emissions
http://www.usatoday.com/tech/news/2011-05-31-carbon-emissions-hit-record_n.htm

Greenhouse Emissions Graphic
http://www.epa.gov/climatechange/emissions/globalghg.html

ANTICIPATION GUIDE

Purpose: To identify what you already know about hydroelectric power, geothermal energy and renewable energy, to direct and personalize your reading, and to provide a record of what new information you learned.

———————◆———————

Before you read "Iceland's Power Struggle," examine each statement below and indicate whether you agree or disagree. Be prepared to discuss your reactions to the statements in groups.

- Iceland is so named because it is covered by a massive sheet of ice.
- Iceland is known for its pristine landscape and commercial fishing, which is the way of life.
- Alcoa aluminum opened a smelting plant in Iceland in 2007 because Iceland is rich in bauxite ore, which is refined to make aluminum.
- Iceland built a 193-meter high dam to accommodate the Alcoa smelting plant.
- Iceland's extremely cheap electricity, absence of "red tape," and minimal environmental impact makes Iceland an attractive place to locate an industrial business.
- Iceland is a "powerhouse" of virtually untapped geothermal and hydroelectric energies.
- The Alcoa refinery in Iceland worked to make the plant "eco-friendly."
- Typical geothermal energy is captured by drilling boreholes 2 kilometers (1.2 miles) deep into the Earth.
- Iceland is exploring methods to drill geothermal boreholes as much as 5 kilometers deep.
- Icelanders are happy that the power company is working more toward using geothermal energy than building more dams for hydroelectric power.

ICELAND'S POWER STRUGGLE

By Marguerite Del Giudice

Two locals kids grin posing for a portrait.
© Sisse Brimberg/National Geographic Stock

THE PEOPLE OF ICELAND AWAKEN TO A STARK CHOICE:

EXPLOIT A WEALTH OF CLEAN ENERGY OR KEEP THEIR LANDSCAPE PRISTINE.

One of the main things to understand about Iceland is how tiny the population is and what it can be like to live here because of that. There's the feeling that everybody on this isolated subarctic island knows just about everybody else, or at least can be associated (through family, friends, neighborhood, profession, political party, or school) by no more than one degree of separation. Imagine a country of 310,000 people, with most of them jammed in and around Reykjavík—a hip European capital known for its dimly lit coffeehouses, live music, and hard-drinking nightlife. That's where all the good jobs are, and the chances of running into somebody you know are so high that it's hard, as one commentator mused, to have a love affair without getting caught.

"We are," said one bespectacled sixtysomething newspaper editor wearing a blazing white shirt, "very close-knit." Then he clasped his hands together, as if in an embrace. Or a vise.

The consequence of living in what amounts to a small town on an island in the middle of nowhere, with its vertiginous links going

It's like living on a mobile—disturbing any part of it could generate a ripple throughout.

back dozens of generations to the origins of Viking myth (a gene pool so pure that molecular biologists drool), is that it functions somewhat like a big extended family. "As soon as you open your mouth," one observer said, "they're all over you." It's like living on a mobile—disturbing any part of it could generate a ripple throughout. So while Iceland in many ways remains an open and transparent society, there's an underlying guardedness among the people when it comes to talking politics and public policy—concerning such things as how the country should go about striking a balance between protecting its environment and growing its economy, which is more or less what this story is about.

The Drowning

In the fall of 2006, secluded, faraway Iceland found itself at a turning point. A remote highland wilderness was being flooded—this to create a reservoir *(Continued on page 22)*

Adapted from "Iceland's Power Struggle" by Margaurite Del Guidice: National Geographic Magazine, March 2008.

Iceland's highest paying jobs and two-thirds of its people are packed in and around Reykjavík, the only city and the center for environmental activism.

(Continued from page 19) measuring 22 square miles as a power source for a new aluminum smelter. The dam that went with the reservoir was the tallest of its kind in Europe (the continent Iceland is conventionally associated with), and the land was going to be irreversibly changed: highland vegetation submerged, waterfalls and part of a dramatic canyon dried up, pink-footed geese and reindeer herds displaced. Environmentalists around the world were condemning the flooding as an attack on one of Europe's last intact wilderness areas—they called it "the drowning"—and the Icelanders themselves didn't know if they were headed for an economic boom, an economic bust, and/or the greatest environmental disaster in European history.

But that's getting ahead of things. This modern Icelandic saga actually begins millions of years ago, for it is rooted in the land itself—the island's unique geology and the geologic destiny that issues from it. First of all, the country is largely uninhabitable—a rocky, windswept, treeless terrain, unsuitable for much of anything beyond raising sheep. "Forbidding" comes to mind. Breathtaking. Strange. Giant chunks of blue ice floating in glacial lakes edged with boiling mud. Craggy mountains with formations that resemble human heads. Volcanoes, geysers, glaciers, belching gas vents, and vast stretches of gnarly lava fields where American astronauts came in the 1960s to see what they would be up against on the moon.

Here's where geologic destiny comes in. Iceland happens to be situated right on top of the intersection of two of Earth's tectonic plates, straddling a volcanic boundary called the Mid-Atlantic Ridge. Consequently, a third of all the lava that has erupted from the Earth in the past 500 years has flowed

Now elsewhere in the world, Iceland may be spoken of, somewhat breathlessly, as western Europe's last pristine wilderness.

out right here, and there are so many natural hot springs that almost all the homes and buildings are heated geothermally. On the surface, meanwhile, sit giant glaciers and the abundant rivers that flow from them. This hot-and-cold combination, of churning activity beneath the surface and powerful rivers above it, makes Iceland one of the most concentrated sources of geothermal and hydroelectric energy on Earth—clean, renewable, green energies that the world increasingly hungers for.

The thing is, very little of that energy has been tapped, because it's stranded in the middle of the nowhere between continental Europe and Greenland. So, since the 1960s, the government has been wooing heavy industry to Iceland with the promise of cheap electricity, no red tape, and minimal environmental impact. But—except for two small smelters and a ferrosilicon plant—getting companies to come here has been a hard sell. The labor force is very small, highly paid, and probably overeducated. Add to that the remoteness of the place, the long, dark winters, and the inhospitable weather. Only an industry requiring the most intensive use of energy, and which could get a heckuva good rate for it over a long period, would find it economical to set up shop all the way in Iceland. The most obvious fit was the aluminum industry. And so it was—to the alarm of environmentalists who want to save that rare land and the thrill of industrialists who want to use some of it to finally produce something—that the paths of aluminum smelting and unspoiled Iceland were fated to cross.

According to Sigurður Arnalds, the spokesman for Landsvirkjun, the national power company—an avuncular engineer everyone calls "Siggi," whose hooded eyes and white-fringed balding head give him the same soft appeal as Mr. Magoo—the grand idea was to "export electrical power on ships in the form of aluminum."

We Have to Live

Now elsewhere in the world, Iceland may be spoken of, somewhat breathlessly, as western Europe's last pristine wilderness. But the environmental awareness that is sweeping the world had bypassed the majority of Icelanders. Certainly they were connected to their land, the way one is complicatedly connected to, or encumbered by, family one can't do anything about. But the truth is, once you're off the beaten paths of the low-lying coastal areas where everyone lives, the roads are few, and they're all bad, so Iceland's natural wonders have been out of reach and unknown even to its own inhabitants. For them the land has always just been there, something that had to be dealt with and, if possible, exploited—the mind-set being one of land as commodity rather than land as, well, priceless art on the scale of the "Mona Lisa."

When the opportunity arose in 2003 for the national power company to enter into a 40-year contract with the American aluminum company Alcoa to supply hydroelectric power for a new smelter, those who had been dreaming of something like this for decades jumped at it and never looked back. Iceland may at the moment be one of the world's richest countries, with a 99 percent literacy rate and long life expectancy. But the project's advocates, some of them getting on in years, were more emotionally attuned to the country's century upon century of want, hardship, and colonial servitude to Denmark, which officially had ended only in 1944 and whose psychological imprint remained relatively fresh. For the longest time, life here had meant little more than a sod hut, dark all winter, cold, no hope, children dying left and right, earthquakes, plagues, starvation, volcanoes erupting and destroying all vegetation and livestock, all spirit—a world revolving almost entirely around the welfare of one's sheep and, later, on how good the cod catch was. In the outlying regions, it still largely does.

Ostensibly, the Alcoa project was intended to save one of these dying regions—the remote and sparsely populated east—where the way of life had steadily declined to a point of desperation and gloom. After fishing quotas were imposed in the early 1980s to protect fish stocks, many individual boat owners sold their allotments or gave them away, fishing rights ended up mostly in the hands of a few companies, and small fishermen were virtually wiped out. Technological advances drained away even more jobs previously done by human hands, and the people were seeing everything they had worked for all their lives turn up worthless and their children move away. With the old way of life doomed, aluminum projects like this one had come to be perceived, wisely or not, as a last chance. "Smelter or death."

The contract with Alcoa would infuse the region with foreign capital, an estimated 400 jobs, and spin-off service industries. It also was a way for Iceland to develop expertise that potentially could be sold to the rest of the world; diversify an economy historically dependent on fish; and, in an appealing display of Icelandic can-do verve, perhaps even protect all of Iceland, once and for all, from the unpredictability of life itself.

"We have to live," Halldór Ásgrímsson said in his sad, sonorous voice. Halldór, a former prime minister and longtime member of Parliament from the region, was a driving force behind the project. "We have a right to live."

The Little Country That Could

At first, most of the country appeared to be behind the dam and smelter that would save the east—it would be good for Iceland, progressive, modern. *(Continued on page 26)*

WILD—FOR HOW LONG?

Only 100,000 people live outside Reykjavík, most in scattered coastal villages that are dying out as family-owned fishing businesses decline. Advocates of aluminum smelting say such development is needed to reinvigorate the rural regions, while environmentalists press for new national parks to support tourism, the fastest growing industry. Last year the government created Vatnajökull National Park (purple) to offset wilderness lost to Kárahnjúkar (orange), though most of it is ice.

MAP BY HANS H. HANSEN, FIXLANDA EHF. SOURCES: ORKUSTOFNUN AND ICELAND GEOSURVEY. M. BRODY DITTEMORE, NGM MAPS

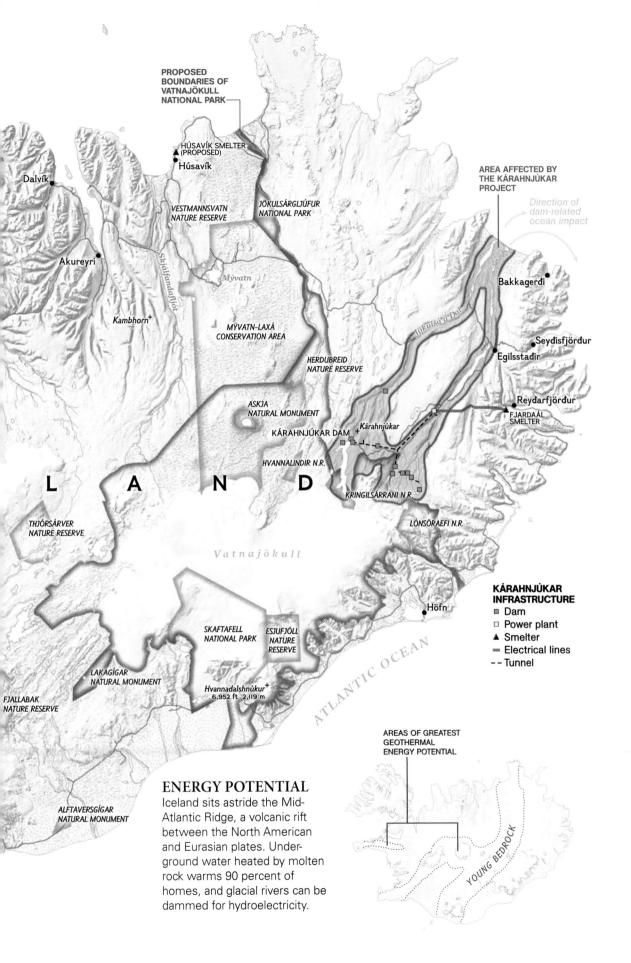

PROPOSED
BOUNDARIES OF
VATNAJÖKULL
NATIONAL PARK

HÚSAVÍK SMELTER
(PROPOSED)
Húsavík

Dalvík

AREA AFFECTED BY
THE KÁRAHNJÚKAR
PROJECT

Direction of
dam-related
ocean impact

VESTMANNSVATN
NATURE RESERVE

JÖKULSÁRGLJÚFUR
NATIONAL PARK

Akureyri

Bakkagerði

Mývatn

Skjálfandafljót

Kambhorn⁺

MÝVATN–LAXÁ
CONSERVATION AREA

Seydisfjördur

Egilsstadir

Jökulsá á Dal

HERDUBREID
NATURE RESERVE

Reydarfjördur

FJARDAÁL
SMELTER

ASKJA
NATURAL MONUMENT

Kárahnjúkar

KÁRAHNJÚKAR DAM

HVANNALINDIR N.R.

L   A   N   D

KRINGILSÁRRANI N.R.

LÓNSÖRAEFI N.R.

THJÓRSÁRVER
NATURE RESERVE

Vatnajökull

Höfn

**KÁRAHNJÚKAR
INFRASTRUCTURE**

▣ Dam
▢ Power plant
▲ Smelter
═ Electrical lines
-- Tunnel

SKAFTAFELL
NATIONAL PARK

ESJUFJÖLL
NATURE
RESERVE

LAKAGÍGAR
NATURAL MONUMENT

FJALLABAK
NATURE RESERVE

Hvannadalshnúkur⁺
6,952 ft 2,119 m

ATLANTIC OCEAN

ALFTAVERSGÍGAR
NATURAL MONUMENT

ENERGY POTENTIAL

Iceland sits astride the Mid-
Atlantic Ridge, a volcanic rift
between the North American
and Eurasian plates. Underground water heated by molten
rock warms 90 percent of
homes, and glacial rivers can be
dammed for hydroelectricity.

AREAS OF GREATEST
GEOTHERMAL
ENERGY POTENTIAL

YOUNG BEDROCK

(Continued from page 23) As far as Icelanders are concerned, Iceland is the greatest country on Earth and everything Icelandic is the best. "We are like Tarzan, a proud island nation," remarked a man who directs a whale museum. They pride themselves on the sagas, the absorbing, rambling, medieval narratives of early Norse and Icelandic society that are generally considered among the Middle Ages' finest literature; a Nobel laureate for literature (Halldór Laxness in 1955); three Miss Worlds; four climbers of Everest; the oldest parliament extant; and the world's first democratically elected female president, Vigdís Finnbogadóttir, who served for 16 years before she'd had enough.

In any event, the national power company was spending 1.5 billion dollars on the hydroelectric part of the dam-and-smelter project, most of it borrowed from international banks—the biggest construction investment little Iceland had ever undertaken and probably ever would. The energy expected to be generated annually (4,600 gigawatt-hours) was about half what the entire nation was then using, and the scale of it was intoxicating: a huge tangle of dams, tunnels, power stations, and high-tension lines, including one rock-and-gravel dam 650 feet high.

All this to service the single aluminum smelter being built on the other side of the country from Reykjavík, in the eastern fjord town of Reyðarfjörður, which is pronounced something like "radar-f'your-dur." There's an unremarkable mountain out there named Kárahnjúkar, and that's where they got the name for what they were doing: the Kárahnjúkar Hydroelectric Project. You say it "kar-en-yoo-kar."

As the project progressed, it gradually became clear that Kárahnjúkar was bigger than anyone had imagined. Even Jóhann

> **A**s the project progressed, it gradually became clear that Kárahnjúkar was bigger than anyone had imagined.

Kröyer, project manager for the dams and tunnels, remarked over dinner at a work-site canteen: "I think maybe people didn't realize how huge this project is."

But as the months passed, a growing and significant minority did realize it, and a kind of national family feud erupted—ostensibly framed around the irreversible impact on the land of the gigantic dam, the blocking of two glacial rivers, and the resultant flooding of the highland wilderness for the reservoir. Iceland had obtained an exemption from the Kyoto Protocol pollution limits, which would expire in 2012, adding an element of urgency, and future smelters and expansions were on the drawing board. Was the government going to take one of the world's cleanest countries and offer it up as a dumping ground for heavy industry?

Did the people really want this—did they even understand what it meant?

Planetary Differences

Kárahnjúkar touched off what one player characterized as Iceland's cold war. A leading conservationist said to me, for instance, that when he walks by a river, he sees an act of God, and that when dambuilders walk by, "they start counting kilowatt-hours." Told of the remark, a smelter manager sniffed, "They don't understand business, and they don't want to."

One side may as well have been from Venus and the other from Mars.

The main guy here from Alcoa, Tómas Már Sigurðsson, a native Icelander with a degree in environmental engineering who considers himself an environmentalist, was upbeat and idealistic. Alcoa's mission, he said, was to be a good neighbor in the community—while creating the most efficient, safe, and eco-friendly smelter on the planet, by recycling materials

Boiling mud pits, a source of alternative energy.
© Ralph Lee Hopkins/National Geographic Stock

and using state-of-the-art technologies to minimize waste and control the sulfur dioxide fumes that are a by-product of smelting aluminum from alumina, a white powder refined from bauxite ore.

He seemed particularly psyched about something Alcoa calls the Sustainability Initiative—under which representatives of diverse interest groups (business, government, the power company, community, church, the environment) devise mutually agreed-on standards for holding Alcoa accountable over time: Did Alcoa raise the region's standard of living, provide the kinds of jobs it said it would, treat the environment as promised? The Sustainability Initiative had been put into effect at the new smelter in Reydarfjördur. "A first in the world," Tómas said—and an industry template for how to approach new communities with the controversial idea of building smelters.

"We're not just building a factory that will produce metal," said Tómas, baby-faced, balding, and fit, with high color, a smear of lips, and an open, earnest manner. "It's a far different level."

Young and idealistic eco-warriors, meanwhile, rolled their eyes in droll disbelief. They dismissed things like the Sustainability Initiative as little more than a capitalist lie designed to manipulate and co-opt an unsuspecting public—a "greenwash," as they put it, which is why, in one highly publicized protest action, a band of them dumped buckets of green-dyed yogurty stuff, called skyr (pronounced skeer), on a gathering of aluminum industrialists (including Alcoa's Tómas), who, not realizing at the time that the skyr was just skyr, were understandably alarmed. "They were trying to say aluminum smelters were eco-friendly!" said Arna Ösp Magnúsardóttir, *(Continued on page 30)*

(Continued on page 30)

A geothermal well field in the Krafla Geothermal Area.

© Ralph Lee Hopkins/National Geographic Stock

Near Lake Myvatn, a local family is having a cookout using the geothermal properties of the area. Food is lowered into a pit in the ground where it is slowly smoked.

© Sisse Brimberg/National Geographic Stock

(Continued from page 27) a local protest organizer who got arrested that day, as she sat at a sidewalk café in Reykjavík, wearing a pink satin pajama top over black tights.

She conceded that she hadn't really thought skyr was going to stop Kárahnjúkar.

"For me, it's to stop future projects and to hassle them enough that they know they won't get away with this again so easily."

Kárahnjúkar's critics argued that surely something else could have been done to help the east without messing so much with the land. Building something on the scale of Kárahnjúkar to solve things, as one man put it, was like "taking a heartbroken kid and giving him a triple-bypass operation."

At the same time, the people in the east said they had tried everything they could think of—business, industry, tourism. "Nothing worked," said Smári Geirsson, a high school history teacher and onetime head of the 16-member association of eastern municipalities. What he said next beat at the heart of Iceland's cold war: "I think the people who are against this are very worried about the land and the reindeer and the birds. But they never want to discuss the needs of the people. Many live in Reykjavík, and they are against it if we move a stone here," he said, pinching his fingers together as if holding one. "But they live in concrete and asphalt. They want to come here in their Jeeps and have a look at the beautiful nature and the people too. And the people must be few and the more strange the better," he said, shaking his bowed head.

This clash of ideals can be traced to the early 20th century, when Iceland was still poor as dirt and romantic poetry was being written about the harnessing of the waterfalls as the

> **W**ith the old way of life doomed, projects like this one had come to be perceived, wisely or not, as a last chance. Smelter or death.

future of Iceland. During and after World War II, the occupation by thousands of British and then American soldiers led to a massive influx of foreign capital and an investment in some fishing trawlers, which boosted the economy for a while. Then over time the fish stocks declined, the herring disappeared, and Iceland became poor again.

By the mid-1960s, according to Styrmir Gunnarsson, the snowy-haired editor of Reykjavík's leading daily, Morgunblaðið, the mood of the country was shifting back to the idea of becoming well-off "by using waterfalls to produce electricity and also by having aluminum factories." So decade after decade, driven by a desire to build a leg other than fish for Iceland to stand on—and "inspired by the idealism of it" and by the poetry, Styrmir said—the society went about creating an infrastructure. Government agencies, ministries, academic departments, financial institutions, engineering firms—all in pursuit of what was thought of as a big and beautiful idea. Then, over the past two decades, the mood of the country again began to shift.

Little by little, Icelanders and eco-trekkers from around the world started venturing into the interior wilderness, he said, and a belated awareness of what was there—"glaciers, black sand, beautiful blue rivers"—began to take hold. "As the population has traveled to these parts, all get the same feeling: You shouldn't change it. No power dams, no roads. It should be just as it is."

In the midst of this environmental awakening, the opportunity with Alcoa stood as possibly the last chance for many Icelanders to realize a century-long industrial dream.

"They felt they were doing something good for the nation," Styrmir said.

Now that they're being depicted as environmental criminals, they're bewildered.

"They don't understand."

The Sky Is Falling

The months passed, and protests mounted. A protest camp and a protest concert were held, featuring Björk, the waif rock chanteuse who is currently Iceland's most famous export; upwards of 10,000 people marched in downtown Reykjavík days before the flooding of the reservoir (the equivalent of ten million-plus people showing up someplace in the U.S.); and state television's Ómar Ragnarsson, Iceland's most famous reporter, a completely bald and preternaturally energetic 67-year-old (known for flying his Cessna to the sites of erupting volcanoes and sleeping in his black Jeep when he had to, surviving on Cheerios and Coca-Cola), launched a 20-foot white fiberglass boat he dubbed The Ark into the reservoir, to collect samples of lost vegetation and stone as it filled over the months and to film the land as it was transformed.

Critics, meanwhile, picked away at Kárahnjúkar's business plan and characterized it, in a variety of ways, as crazy. The borrowed 1.5 billion dollars, for instance, was to be paid off with revenue from Alcoa over the four-decade contract. After that, Siggi, the power company spokesman, predicted, the dam would be "a gold mine"—a rosy forecast not shared by those who had started to think twice about there being no direct return for 40 years on such a huge investment.

And what about the geologic risks associated with boring and blasting 45 miles of tunnels in a country that was one big volcano? Siggi insisted that the dam wasn't located in an earthquake zone but nonetheless had been designed to withstand heavy shocks, and he likened all the safety measures taken to "putting many, many belts on the same pair of trousers."

But would the electromagnetic fields emanating from the project's 31 miles of high-tension power lines make people sick? And how about pollution from the smelter? Alcoa's plan was to blast potentially acid-rain-producing sulfur dioxide fumes way out into the atmosphere through a giant chimney. But what if the fumes became caught in the fjord's notoriously calm atmosphere, trapped between mountains?

Inevitably, it started to dawn on more and more people: The entire public was paying for this grand project and would be responsible if something went wrong, but the benefits were accruing only to one depressed region numbering several thousand people. There began to be grumblings about Iceland's "old boy" political party system and the perceived lack of forthrightness in engaging the electorate about the country's industrial policy.

"On the surface, we have an open debate," said Baldur Þórhallsson, a professor at the University of Iceland who specializes in the politics of small societies. "But underneath, there is this tendency by some politicians to control the debate and the agenda... phone calls, emails, and letters making little indirect threats." With job opportunities few in such a small place—and ties among politics, business, government, and media so Byzantine that conflicts of interest are impossible to avoid and sleeping with the enemy is literally true—it's understood that speaking one's mind could be a really bad career move.

Most people I talked to seemed wary of appearing to demean their relatives to a stranger and sensitive about giving a bad impression of Iceland. But once they got going, it was talk, talk, talk into the night, spewing every story and conspiracy theory imaginable—most of them fluent in English and otherwise speaking an ancient Viking tongue that sounded like a tape recorder playing backward. There were even glimpses of an abiding belief in (Continued on page 34)

A tourist observes boiling mud pits, a source of alternative energy.
© Ralph Lee Hopkins/National Geographic Stock

(Continued from page 31) elves and the "hidden people," who live in rocks that roads have been routed around so as not to disturb them. What kind of spiritual havoc might come from moving the Earth on the scale of Kárahnjúkar?

The Problem of Choice

Late one night, in a hut on a lake deep in the eastern highlands, Ómar, the reporter, while empathizing by candlelight with the people of the regions, lamented the larger course of events. "If only instead of looking for ways to use the land for heavy industry, Iceland had kept the land intact and for 20 years had worked on how to sell her intact, we'd have gotten much more for her."

Why wasn't Iceland positioning itself as a world leader in the development of hydrogen alternatives? Or marketing what is arguably the world's oldest parliamentary democracy as a mecca for law students? Or more aggressively branding Icelandic products in the international market, which would surely pay for such remote purity? How about some high-end, eco-conscious tourism on a par with, say, the Galápagos Islands? Or capitalizing on the country's rich heritage?

At the Culture House, for instance, you can view original medieval manuscripts, like Egil's Saga, smudgy pages in careful script written on calfskin vellum. The Codex Regius of the Edda poems is also there, the oldest and most important collection of poems describing the gods, heroes, and mythology of antiquity, said to have inspired artists from Wagner to Tolkien.

"But we haven't sold this," said Andri Snær Magnason, author of Dreamland: Self-Help for a Frightened Nation, a surprise, runaway best-seller that exposed the spell he thought the Icelandic people had been under regarding the dam and smelter and galvanized the conservation movement. "We're selling people cuddly stuffed puffins but not promoting real depth."

Even more important, modern Icelandic society, Andri felt, had no template for how to conduct a national dialogue. As a result, the majority of people, instead of making conscious, informed decisions about issues like Kárahnjúkar, had—out of a combined sense of trust in the system, fear of the system, and intellectual laziness in the face of complicated and precarious issues—passively supported something they didn't really understand.

Here was Iceland (or at least Reykjavík), Andri said, "at a peak of economic growth and progress"—thanks to thriving pharmaceutical and tech industries, the privatization of the banks, and a general liberalization of the financial sector over the past 20 years. At such a juncture, he said, the problem should be "a problem of choice"—not of smelter or death. Not the old fear of privation but a sense that the future is limited only by Icelanders' belief in themselves and their willingness to reawaken that primeval Viking spirit idling restlessly behind their seeming impassivity: the daring and inventiveness, the sense of artistry, and of tremendous resolve, from the old ways and the old days of old Iceland, when it was believed, as one former egg farmer told me, "that you could change the world with a poem."

So here they all were, one big unhappy family. The environmentalists were trying to save Iceland. The industrialists were trying to save Iceland. Everyone was trying to save Iceland.

"A mental civil war," somebody called it.

A war of dreams.

What Price Freedom?

Soon after the flooding, in what looked like a concession to the burgeoning green constituency, the government decided to designate a huge, long-awaited national park—a wilderness area in and around Vatnajökull glacier, adjacent to Kárahnjúkar, that may encompass some 5,000 square miles.

The park was a major victory, perhaps the first of many, it was hoped by conservation leaders such as Árni Finnsson, director of the Iceland Nature Conservation Association, a restless, sensitive man with a flop of strawberry blond hair and a voice low and raspy from his having lost one of his vocal cords in a car accident. He was ecstatic about the park and also the movement's enhanced political clout. A new, green party emerged, chaired by Ómar, called Iceland's Movement; and though it failed in last May's elections to garner the minimum 5 percent of the vote required by law to earn any seats in the parliament (it got 3.3 percent), and its future was unclear, the political landscape appeared to be changing. The greens were squarely in the game now, environmental issues were higher on the public agenda, and all of Iceland's parties were going to have to take them into account. Pro-Kárahnjúkar people were saying things like "we're nearing the end of the road on aluminum," and energy-intensive alternatives such as "server farms," huge computer-processing centers, were on the horizon, as was a shift to geothermal energy, which can be less invasive than hydro.

Meanwhile, on the aluminum side, as of last fall: Smelting had begun at the Alcoa plant in Reyðarfjörður, with full production expected this year—344,000 metric tons annually; the company was planning another smelter, using geothermal energy, in the northeast town of Húsavík; and Century Aluminum was expected to start building a second smelter this year, near the airport outside Reykjavík. (Though the townspeople of Hafnarfjörður narrowly voted down Alcan's bid to increase the size of its existing smelter there.)

So for now, Iceland is in a state that can only be described as schizophrenic.

"The question over the next few years," said Styrmir, the newspaper editor, "is can we compromise between those who are still inspired by the idea of Iceland becoming rich by using our land for energy and the people who are fighting against any changes in the environment? Can we use geothermal energy in a better way than how it's been done in the past, with ugly above-ground zigzagging pipes that carry the steam and lie across the landscape like a broken zipper?"

A new, deep-drilling technology that could generate more power with fewer boreholes and exposed pipes is being explored by the national power industry (partly supported by Alcoa). The experiments involve drilling five kilometers down instead of the usual two, said Landsvirkjun's Siggi, to tap into "higher temperature and higher pressure and therefore up to ten times more power per borehole." But the results are years away. "Such a hot liquid is not easy to handle, so there are many technical hurdles." At the same time, the use of conventional geothermal technology, mostly to power aluminum plants in the southwest and northeast, is also being developed—and challenged by environmentalists. "Most of this is still in the exploration phase," said Siggi, about which Ómar remarked: "They call it research, but for that research they've got bulldozers in there and made roads and destroyed the area."

So the cold war slogs on, with geothermal energy emerging as the next environmental battleground.

For Iceland, and for any country groping for a balance between protecting its environment and growing its economy, the challenge is this: How do you change the infrastructure of a society—overcome the inertia and allow time for alternatives to develop—while people's lives and livelihoods are at stake? How do you develop eco-friendly industries that are also lucrative, thereby making conservation profitable as well as virtuous?

"I think Kárahnjúkar is the beginning of a new era," said a man filming a documentary about it. "An opportunity for transformation."

Career Investigation

In a major news announcement today, it was announced that there was only enough crude oil reserves in the entire world to last ten years at the most. We knew this day was coming, but now there is a hard deadline. We must look at alternative methods of providing energy to create electricity and heat, fuel our cars, etc. Who will create those alternative energies? Possibly YOU!

Listed here are several possible careers to investigate. You may also be able to find a career path not listed. Choose three possible career paths and investigate what you would need to do to be ready to fill one of those positions.

Among the items to look for:

- Education—Is a college degree necessary? What would you major in? Should you have a minor?

- Working conditions—What is the day to day job like? Will you have to be out in the field all day, or is it a desk job? Is the work strenuous? Are you working on a drilling rig at sea, away from your family for 3–6 months at a time? Is the job "hazardous"?

- Pay scale—What is the average "starting pay" for the position you are seeking? Don't be fooled by looking at "average pay" which may include those who have been working for 20+ years.

- Is the job located in the states, or is there a possibility to travel to other parts of the world?

- Are there any other special requirements?

Architect
Architectural Technician or Technologist

Biologist
Botanist
Chemical Engineer
Chemical Engineering Technician
Chemist
Civil Engineer
Civil Engineering Technician
Ecologist
Electrician
Electricity Distribution Worker
Engineering Construction Technician
Geoscientist
Geotechnician
Hydrogeologist
Materials Engineer
Microbiologist
Plumber
Sheet Metal Worker
Steel Erector
Structural Engineer
Sustainability Director
Town Planner
Town Planning Support Staff
Welder

Team Building Activity

You will be assigned to teams of four or five students and host a panel that is composed of experts from around the world in the field of green technologies. The panel will examine the impact of the Kárahnjúkar Hydroelectric Project as well as the Alcoa smelting plant in Iceland. The team will use PowerPoint, posters, videos, and other supporting resources to solidify their points. Each team member should cover a specific category (some examples are listed) to solidify the position of the group either for or against additional construction.

After the presentation, the audience (representing members of the media and/

or experts from the field) will be allowed to question the panel.

- The Alcoa smelting plant and its effects on the economy and land.
- Geothermal energy (both heat and power generation).
- Hydroelectricity.
- Governmental agencies and their perspective on the refinery and development of the land.
- Members of the public at large and their perspective on the refinery and development of the land.

Writing Assignment

After reading the article "Iceland's Power Struggle" and conducting your own research, write a brief paper discussing Iceland's attempt to balance economic development while maintaining its natural resources and once pristine landscape.

In addition to your own ideas and thoughts, please discuss:

- The basics of geothermal energy.
- The basics of hydroelectric power.
- Why Iceland is rich in "green" energies.
- Why Iceland allowed the Kárahnjúkar Hydroelectric Project to proceed.

Listed here are a few links to help you in getting started.

Links:

Karahnjukar Hydroelectric Project
http://www.mwhglobal.com/mwh-projects/karahnjukar-hydroelectric-project

Three Robbins TBMs Carve Out Hydroelectric Tunnels in Iceland
http://www.robbinstbm.com/case-study/karahnjukar-hydropower-project/

Alcoa Mining 1
http://www.alcoa.com/iceland/en/alcoa_iceland/making_mining.asp

Alcoa Ore Refining 1
http://www.alcoa.com/iceland/en/alcoa_iceland/refining.asp

Saving Iceland
http://www.savingiceland.org/tag/alcoa/

Iceland Divided Over Aluminum's Role in Its Future
http://articles.latimes.com/2011/mar/26/business/la-fi-iceland-economy-20110326

"Century Aluminum" Tag Archive
http://www.savingiceland.org/tag/century-aluminum/

Iceland's Environment Minister Discusses Geothermal Power, Aluminum Concerns
http://green.blogs.nytimes.com/2009/07/08/icelands-environment-minister-discusses-geothermal-power-aluminum-concerns/

Purpose: To identify what you already know about solar energy, to direct and personalize your reading, and to provide a record of what new information you learned.

◆

Before you read "Plugging into the Sun," examine each statement below and indicate whether you agree or disagree. Be prepared to discuss your reactions to the statements in groups.

- Nevada is home to an array of curved mirrors covering more than 250 acres.
- The mirrors at the Nevada Solar One site are curved because the sun is so hot that the mirrors have semi melted into the curved shape.
- At the Nevada Solar One site, collected sunrays, heat, and water to create steam.
- Solar energy plants such as Nevada Solar One are very efficient.
- Solar energy, being a green energy, cannot produce enough energy to be beneficial.
- Geothermal and wind energy are better sources of renewable energy than solar energy.
- At 250 acres and 182,000 mirrors, Nevada Solar One is the largest solar power generating site in the United States.
- Typically solar energy is used to heat water which will turn into steam at a generating unit, but recently a Spanish company has been working with heating molten salt and storing heat energy that is then used to heat water to produce steam for the generating unit.
- Rather than using solar power to heat water to convert to steam, using photovoltaic panels would be more efficient.
- Photovoltaic solar power efficiency is only about 10–20 percent.
- Some photovoltaic panels can be applied to walls with a paint roller.
- To make photovoltaic solar energy realistic, energy must be stored for use when there is no sunlight.
- To make solar energy "pay off" requires many hundreds of acres of mirrored panels.

PLUGGING INTO THE SUN

By George Johnson

Photographs by Michael Melford

At an electric plant in southern Spain, mirrors as big as houses catch some of the 120 quadrillion watts of sunlight that constantly fall on Earth. Government subsidies for this pricey yet promising power source have made Europe the world's solar capital.

Turning unused space into a power source for 1,300 homes, Southern California Edison contractors cover a 14-acre warehouse roof near Los Angeles with some 33,000 lightweight photovoltaic, or PV, panels. California law requires utilities to generate 20 percent of their power from renewable sources by 2010.

SUNLIGHT BATHES US

IN FAR MORE ENERGY THAN WE COULD EVER NEED–

IF WE COULD JUST CATCH ENOUGH.

The optimists say **solar power could become as economical and efficient as fossil fuels. The pessimists say they've heard all this before.**

Early on a clear November morning in the Mojave Desert, the sun is barely touching the peaks of the McCullough Range with a cool pink glow. Behind them, a full moon is sinking over the gigawatt glare of Las Vegas. Nevada Solar One is sleeping. But the day's work is about to begin.

It is hard to imagine that a power plant could be so beautiful: 250 acres of gently curved mirrors lined up in long troughs like canals of light. Parked facing the ground overnight, they are starting to awaken—more than 182,000 of them—and follow the sun.

"Looks like this will be a 700-degree day," says one of the operators in the control room. His job is to monitor the rows of parabolically shaped mirrors as they concentrate sunlight on long steel pipes filled with circulating oil, heating it as high as 750 degrees Fahrenheit. From the mirror field, the blistering liquid pours into giant radiators that extract the heat and boil water into steam. The steam drives a turbine and dynamo, pushing as much as 64 megawatts onto the grid—enough to electrify 14,000 households or a few Las Vegas casinos. "Once the system makes steam, it's very traditional—industry-standard stuff," says plant manager Robert Cable, pointing toward a gas-fired power plant on the other side of Eldorado Valley Drive. "We get the same tools and the same parts as the place across the street."

When Nevada Solar One came on line in 2007, it was the first large solar plant to be built in the U.S. in more than 17 years. During that time, solar technology blossomed elsewhere. Nevada Solar One belongs to Acciona, a Spanish company that generates electricity here and sells it to NV Energy, the regional utility. The mirrors were made in Germany.

Putting on hard hats and dark glasses, Cable and I get into his pickup and drive slowly past row after row of mirrors. Men with a

Adapted from "Plugging into the Sun" by George Johnson: National Geographic Magazine, September 2009.

water truck are hosing down some. "Any kind of dust affects them," Cable says. At the far edge of the mirror field, we stop and step out of the truck for a closer look. To show how sturdy the glass is, Cable bangs it like a drum. Above his head, at the focal point of the parabola, the pipe carrying the oil is coated with black ceramic to soak up the light, and it's encased in an airless glass cylinder for insulation. On a clear summer day with the sun directly overhead, Nevada Solar One can convert about 21 percent of the sun's rays into electricity. Gas plants are more efficient, but this fuel is free. And it doesn't emit planet-warming carbon dioxide.

About every 30 seconds there is a soft buzz as a motor moves the mirrors a little higher; by midday they will be looking straight up. It's so quiet out here one can hardly fathom how much work is being done: Each of the 760 arrays of mirrors can produce about 84,000 watts—almost 113 horsepower. By 8 a.m. the oil coursing through the pipes has reached operating temperature. A white plume is spewing from a cooling stack. Half an hour later, the sound of the turbine inside the generating station has reached a high-pitched scream. Nevada Solar One is ready to go on line.

With a new administration in Washington, D.C., promising to take on global warming and loosen the grip of foreign oil, solar energy finally may be coming of age. Last year oil prices spiked to more than $140 a barrel before plunging along with the economy—a reminder of the dangers of tying the future to something as unpredictable as oil. Washington, D.C., confronting the worst recession since the 1930s, is underwriting massive projects to overhaul the country's infrastructure, including its energy supply. In his inaugural address President Barack Obama promised to "harness the sun and the winds and the soil

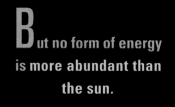

But no form of energy is more abundant than the sun.

to fuel our cars and run our factories." His 2010 budget called for doubling the country's renewable energy capacity in three years. Wind turbines and biofuels will be important contributors. But no form of energy is more abundant than the sun.

"If we talk about geothermal or wind, all these other sources of renewable energy are limited in their quantity," Eicke Weber, director of the Fraunhofer Institute for Solar Energy Systems, in Freiburg, Germany, told me last fall. "The total power needs of the humans on Earth is approximately 16 terawatts," he said. (A terawatt is a trillion watts.) "In the year 2020 it is expected to grow to 20 terawatts. The sunshine on the solid part of the Earth is 120,000 terawatts. From this perspective, energy from the sun is virtually unlimited."

There are two main ways to harness it. The first is to produce steam, either with parabolic troughs like the ones in Nevada or with a field of flat, computer-guided mirrors, called heliostats, that focus sunlight on a receiver on top of an enormous "power tower." The second way is to convert sunlight directly into electricity with photovoltaic (PV) panels made of semiconductors such as silicon.

Each approach has its advantages. Right now steam generation, also known as concentrating solar or solar thermal, is more efficient than photovoltaic—a greater percentage of incoming sunlight is converted into electricity. But it requires acres of land and long transmission lines to bring the power to market. Photovoltaic panels can be placed on rooftops at the point where the power is needed. Both energy sources share an obvious drawback: They fade when it's cloudy and disappear at night. But engineers are already developing systems for storing the energy for use in the darker hours. *(Continued on page 44)*

As the lights come on in Los Angeles, solar evangelist Larry Kazmerski models the latest in PV: bendy thin-film panels that fit so many places, says the National Renewable Energy Laboratory researcher, they could one day power whole cities. "This is going to be on every roof and building," Kazmerski says.

The energy of the future has a past. Shards of old mirrors lie under their replacements at California's 25-year-old SEGS I, the first commercial solar thermal plant in the U.S.

(Continued from page 42) The optimists say that with steady, incremental improvements—no huge breakthroughs are required—and with substantial government support, solar power could become as economical and efficient as fossil fuels. The pessimists say they've heard all this before—30 years ago, during the presidency of Jimmy Carter. That too was a period of national crisis, triggered by the Arab oil embargo of 1973. Addressing the nation in his cardigan sweater, President Carter called for a new national energy policy with solar energy playing a large part. In 1979 the Islamic revolution in Iran sent oil prices soaring again. American drivers lined up for gasoline, their radios blaring songs like "Bomb Iran," by Vince Vance and the Valiants (sung to the tune of the Beach Boys' "Barbara Ann"). Carter,

true to his word, put solar water heaters on the White House roof.

During the next few years, two large fields of parabolic troughs, SEGS I and II (for Solar Electric Generating Station) were installed about 160 miles southwest of Las Vegas, near Daggett, California. They were followed by seven more plants nearby, at Kramer Junction and beside waterless Harper Lake. The plants are still operating—about a million mirrors in all on some 1,600 acres with a combined power of 354 megawatts. From afar they look like mirages.

The momentum didn't last. As the economy adjusted to the Iranian oil shock, fuel prices fell. With the sense of urgency reduced, along with the research dollars, solar remained a minor factor in the energy equation. The SEGS plants were still being built when President Ronald Reagan took the solar water heaters off the White House roof. The first solar revolution fizzled.

Two decades later, a new solar revolution may be ready to begin.

Another legacy of the Carter era, the National Renewable Energy Laboratory (NREL) in Golden, Colorado—the government's primary research center for solar, wind, hydrogen, and other alternative fuels—is bracing for a resurgence. When I visited last fall, a new research campus and headquarters were under construction against the side of a mountain outside Golden. Five acres of photovoltaic panels on top of the mesa will feed the labs and offices below. That may be just the beginning. Once treated by the government as something of a stepchild, NREL is benefiting from the extra money the Obama Administration is devoting to renewable energy. "Right now solar is such a small fraction of U.S. electricity production that it's measured in tenths of a percent," said Robert Hawsey, an associate director of the lab. "But that's expected to grow. Ten to 20 percent of the nation's peak electricity demand could be provided by solar energy by 2030."

But not without government help. Nevada Solar One would never have been built if the state had not set a deadline requiring utilities to generate 20 percent of their power from renewable sources by 2015. (More than two dozen states now have "renewable-portfolio standards," and earlier this year Congress was debating a federal one.) During peak demand—a hot afternoon in Las Vegas, when production costs are highest—the solar plant's electricity is almost as cheap as that of its gas-fired neighbor. But that's only because a 30 percent federal tax credit helped offset its construction costs.

If photovoltaic panels covered just three-tenths of a percent of the United States, a 100-by-100-mile square, they could power the entire country.

Aiming to bring down costs and reduce the need for incentives, NREL's engineers are studying mirrors made from lightweight polymers instead of glass and receiving tubes that will absorb more sunlight and lose less heat. They're also working on solar power's biggest problem: how to store some of the heat produced during daylight hours for release later on. "In the Southwest particularly, peak loads are in the daytime, but they don't end when the sun goes down," said Mark Mehos, an NREL program manager. People come home from work, turn on lights and air conditioners. Before long they may be plugging in electric cars.

Last year the first commercial solar plant with heat storage opened near Guadix, Spain, east of Granada. During the day, sunlight from a mirror field is used to heat molten salt. In the evening, as the salt cools, it gives back heat to make more steam. In Arizona the Solana Generating Station will also use molten salt for storage. When it goes on line in 2012, three square miles of parabolic troughs will produce 280 megawatts for Phoenix and Tucson. Solana is being built by a Spanish company, Abengoa Solar—an indication of just how far, in the development of this technology, the United States has fallen behind.

Back in the 1980s, an engineer named Roland Hulstrom calculated that if photovoltaic panels—the other big solar technology—covered just three-tenths of a percent of the United States, a 100-by-100-mile square, they could electrify the entire country.

People thought he wanted to pave the Mojave with silicon. "The environmentalists got up in arms and said, You can't just go out and cover a hundred miles square," Hulstrom said recently as he sat in his office at NREL. But that's not what (Continued on page 50)

SunCatchers at Sandia National Laboratories in New Mexico stand dormant at moonrise. At daybreak each mirror array will turn, and concentrated light will heat a Stirling engine held at the focal point, driving pistons and making electricity. No system is more efficient at converting photons to grid-ready AC power. Stirling Energy Systems plans to erect some 60,000 SunCatchers at desert sites near Los Angeles and San Diego.

ENDLESS POTENTIAL

Energy Equal to 6,000 times the world's electricity use constantly shines on Earth. Even with current technology, we could harvest enough to supply dozens of times our demand for electricity—but building the infrastructure needed to switch to solar would cost much more at current prices than continuing to burn fossil fuels. As data from NASA satellites show (map, right), the world's solar leaders, notably Germany, are not the sunniest countries but the ones that can afford to pay extra for solar power. Solar costs are falling steadily, however. Developing nations in the subtropics may benefit from that trend; steady sunshine there means investment in solar infrastructure could pay off fast. Most of the world's best solar potential remains unexploited.

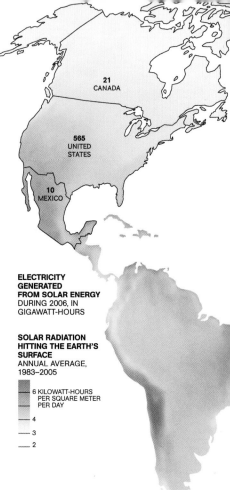

21 CANADA

565 UNITED STATES

10 MEXICO

ELECTRICITY GENERATED FROM SOLAR ENERGY DURING 2006, IN GIGAWATT-HOURS

SOLAR RADIATION HITTING THE EARTH'S SURFACE ANNUAL AVERAGE, 1983–2005

6 KILOWATT-HOURS PER SQUARE METER PER DAY
4
3
2

Electricity that could be generated worldwide from renewable sources
975,010 TERAWATT-HOURS

Electricity generated worldwide in 2006
19,015 TERAWATT-HOURS

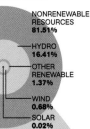

470,278 TWh SOLAR PHOTOVOLTAIC

275,556 TWh CONCENTRATING SOLAR

105,278 TWh WIND (LAND BASED)

91,398 TWh OCEAN (TIDAL AND WAVE)

13,889 TWh HYDRO

12,500 TWh GEOTHERMAL

6,111 TWh WIND (OFFSHORE)

NONRENEWABLE RESOURCES 81.51%

HYDRO 16.41%

OTHER RENEWABLE 1.37%

WIND 0.68%

SOLAR 0.02%

◀ RANKING THE RENEWABLES

The sun's potential for power generation eclipses that of all other renewable energy sources. But for now solar power barely registers in the world's energy portfolio (inset graph). It accounts for only a small fraction of a percent of total electrical output—much less than hydropower or wind energy, which are cheaper to produce.

SEAN MCNAUGHTON, NG STAFF GRAPHIC BY 5W INFOGRAPHICS

SOURCES: NASA (WORLD MAP); WORLD ENERGY STATISTICS AND BALANCES OECD/IEA, 2008; NATIONAL RENEWABLE ENERGY LABORATORY (SOUTHWEST MAP); ECOFYS (POTENTIAL GENERATION)

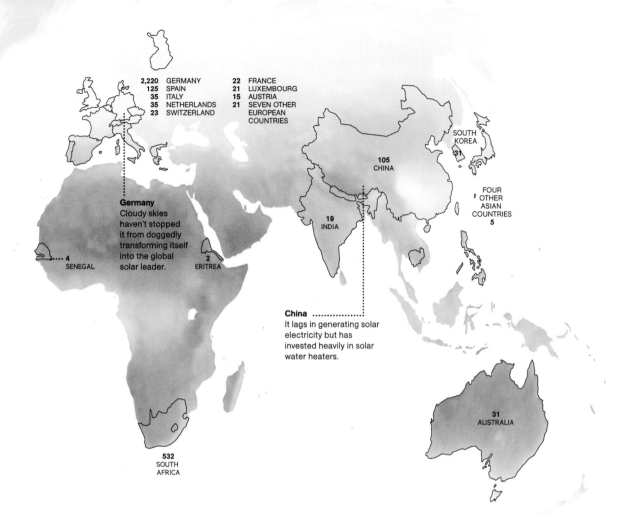

2,220	GERMANY	22	FRANCE
125	SPAIN	21	LUXEMBOURG
35	ITALY	15	AUSTRIA
35	NETHERLANDS	21	SEVEN OTHER
23	SWITZERLAND		EUROPEAN
			COUNTRIES

Germany
Cloudy skies haven't stopped it from doggedly transforming itself into the global solar leader.

4 SENEGAL

2 ERITREA

105 CHINA

19 INDIA

SOUTH KOREA 31

FOUR OTHER ASIAN COUNTRIES 5

China
It lags in generating solar electricity but has invested heavily in solar water heaters.

31 AUSTRALIA

532 SOUTH AFRICA

THE SOLAR SOUTHWEST

Solar advocates say the desert Southwest could light the entire U.S. Thousands of square miles there (colored areas) are not only sunny but also flat and undeveloped enough for concentrating-solar plants. Some environmentalists oppose the land- and water-hungry projects and the new transmission lines that would be needed.

SOLAR RADIATION SUITABLE FOR USE IN CONCENTRATING-SOLAR PLANTS

8 KILOWATT-HOURS PER SQUARE METER PER DAY
7
6
POPULATED AREAS

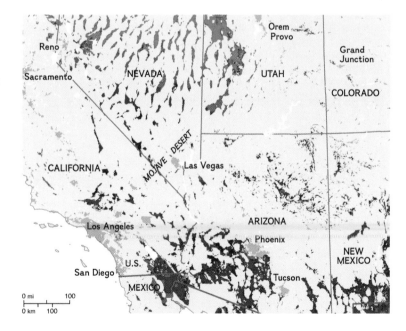

Reno
Sacramento
NEVADA
UTAH
Orem Provo
Grand Junction
COLORADO
CALIFORNIA
MOJAVE DESERT
Las Vegas
Los Angeles
ARIZONA
Phoenix
NEW MEXICO
San Diego
U.S.
MEXICO
Tucson

0 mi 100
0 km 100

(Continued from page 45) he meant. "You can cover parking lots with photovoltaic. You can put it on house roofs."

Twenty years later, PV panels still contribute only a tiny amount to the nation's electricity supply. But on rooftops in California, Nevada, and other states with good sunshine and tax incentives, they're a sight almost as familiar as air conditioners—and though not yet as developed as solar thermal, they may have a brighter future.

Right now the panels are expensive, and they provide an efficiency of only about 10 to 20 percent, compared with the 24 percent of parabolic troughs. History more than physics is to blame. After the solar bust in the mid-1980s, many of the best engineers migrated to the computing industry, which uses the same raw material—silicon and other semiconductors. Following what is called Moore's law, microprocessors doubled in capability every couple of years, while solar languished. Now some of the engineering talent is moving back to solar.

Researchers at NREL are exploiting the fact that different semiconductors capture different colors from a beam of sunlight. By layering compounds called gallium indium phosphide and gallium indium arsenide and using a lens to concentrate sunlight, they built a PV cell last year that is 40.8 percent efficient (a world record, since broken). But it's far from ready for mass production. "The technology is incredibly sophisticated," said Ray Stults, an associate director of the laboratory. "We can make it right now for $10,000 per square centimeter, but not many people are going to buy it."

Another approach is to trade higher efficiency for lower cost. Though they generate less power per square inch, thin-film semiconductors require less raw material, making them a cheaper alternative for large photovoltaic installations. Two American companies, First Solar and Nanosolar, say they can now manufacture thin-film solar cells at a cost of around a dollar a watt—tantalizingly close to what's needed to compete with fossil fuels. Looking further ahead, engineers at NREL are working on photovoltaic liquids. "The goal there is to make it the cost of a gallon of paint," Stults said. "The efficiencies won't be 40 or 50 percent. They'll be 10 percent. But if it's cheap, you can paint your walls, hook it up, and go."

Photovoltaic panels aren't limited to individual houses or warehouses. On the northeastern outskirts of Las Vegas, Nellis Air Force Base is supplying an average of 25 percent of its electricity with photovoltaic. On some winter days when there is no need for air-conditioning, 100 percent of the base is solar powered. Last fall, as I looked across the field of 72,416 sun-tracking panels, the wind blowing between the rows, I could see the appeal: There were no oil pipes, heat exchangers, boilers, dynamos, or cooling towers—just solar photons knocking electrons off silicon atoms and generating a current. Constructed in just 26 weeks in 2007 by the SunPower Corporation, the system generates 14.2 megawatts, making it the largest photovoltaic installation in the United States—though only about the 25th largest in the world. Nearly all the bigger ones are in Spain, which, like Germany, has invested heavily in solar power.

None of those plants yet include a storage system. Since photovoltaics produce electricity directly, there is no heat to capture in tanks of molten salt. One option would be to divert some of the photovoltaic current during the day to drive pumps, compressing air into underground caverns. Compressed air has been employed for decades in Germany and Alabama to store the cheaper nighttime output of conventional power plants for use during the daytime peak. At a solar plant the cycle would be reversed: When electricity was needed at night, the pent-up energy from the sunlit hours would be released, rushing forth and spinning a turbine.

Right now people who live off-grid with PV panels on their roofs rely on ordinary batteries to get through the night. In the future they might have solar-powered electrolyzers that split water molecules into hydrogen and oxygen. Recombining the gases in a fuel cell would yield electricity again. The idea is old, but last year Daniel Nocera, a chemist at MIT, reported what may be a breakthrough: a new catalyst that makes splitting water much cheaper. At public lectures Nocera likes to hold up a large plastic water bottle. All of a family's nighttime electricity requirements, he says, could be stored in five of these, with enough left over to run the electric car.

No one knows in detail the future of solar energy. But there is a gathering sense that it is wide open—if we can make the commitment to jump-start the technology. "Originally it seemed like a pie-in-the-sky idea," Michelle Price, the energy manager at Nellis, told me last fall when I toured the base's new photovoltaic plant. "It didn't seem possible." Many things seem possible now.

On a cold December morning west of Frankfurt, Germany, fog hung frozen in the trees, and clouds blocked the sun. Shivering on a ridge above the town of Morbach, I watched the blades of a 330-foot-high wind turbine swoop in and out of the gloom. Down below, a field of photovoltaic panels struggled for light. Who would have thought that Germany would transform itself into the largest producer of photovoltaic power in the world, with a capacity of more than five gigawatts?

A fraction of this power comes from centralized plants like the small one at Morbach or even the sprawling 272-acre Waldpolenz Solar Park, which was constructed recently with thin-film technology on an abandoned

On some days 100 percent of Nellis Air Force Base is solar powered. No pipes, boilers, or dynamos—just photons knocking electrons off silicon.

Soviet air base near Leipzig. With land at a premium in Germany, solar panels are mounted on rooftops, farmhouses, even on soccer stadiums and along the autobahn. Though dispersed across the countryside, they are connected to the national grid, and utility companies are required to pay even the smallest producers a premium of about 50 euro cents a kilowatt-hour.

"We are being paid for living in this house," said Wolfgang Schnürer, a resident of Solarsiedlung—"solar settlement"—a condominium complex in Freiburg. Outside, snow was sliding off the solar panels that covered the roofs of the development. The day before, Schnürer's system had produced only 5.8 kilowatt-hours, not enough even for a German household. But on a sunny day in May it had yielded more than seven times that much.

After serving coffee and Christmas cookies, Schnürer spread some printouts on the table. In 2008 his personal power plant generated 6,187 kilowatt-hours, more than double what the Schnürers consumed. When the amount they used was subtracted from the amount they produced, they came out more than 2,500 euros (nearly $3,700) ahead.

Sitting at the edge of the Black Forest in the southern part of the country, "sunny Freiburg," as the tourist brochures call it, has been transformed by the solar boom. Across the street from Solarsiedlung, a parking garage and a school are covered with photovoltaic panels. In the older part of town, towering walls of photovoltaics greet visitors at the train station. Nearby, at the Fraunhofer Institute for Solar Energy Systems, the next generation of technology is being developed. In one project, Fresnel lenses are used to

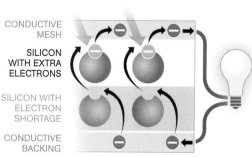

CONDUCTIVE
MESH

SILICON
WITH EXTRA
ELECTRONS

SILICON WITH
ELECTRON
SHORTAGE

CONDUCTIVE
BACKING

Graphic: 5W INFOGRAPHICS

PHOTOVOLTAIC POWER

Solar panels like these on roofs at a Bavarian farm produce electricity when light jars electrons loose in a semiconductor, often silicon (above). Unlike concentrating solar, the other strategy for generating solar electricity, PV systems can operate efficiently on a small scale.

concentrate sunlight 500 times, raising the efficiency of a standard photovoltaic panel as high as 23 percent.

It is the demand created by the government's "feed-in tariff" that drives research like this, said Eicke Weber, the institute's director. Anybody who installs a photovoltaic system is guaranteed above-market rates for 20 years—the equivalent of an 8 percent annual return on the initial investment. "It is an ingenious mechanism," Weber said. "I always say the United States addresses the idealists, those who want to save the planet— the Birkenstock crowd. In Germany the law addresses anyone who wants to get 8 percent return on his investment for 20 years."

The most spectacular showcase for the future of solar is probably Plataforma Solúcar, a Spanish solar energy complex on the Andalusian plains. I'd seen photographs of the 11-megawatt power tower called PS10. Rising 377 feet high, it is surrounded by 624 sun-tracking mirrors that reflect light beams toward its crown, igniting a glow that shines like a new star. Next to it, PS20 has since been completed with twice as many heliostats and double the power. But as I crested a hilltop about 15 miles west of Seville, I saw that the German weather had followed me. The valley was enveloped in fog—a reminder that even in torrid southern Spain, solar *(Continued on page 55)*

Above a scorched plain outside Seville, Spain, reflected sunlight reflects again off low clouds. Ordinarily the mirrors at Abengoa Solar's PS10 station beam searing, concentrated light to the top of the "power tower," heating a boiler that makes steam to drive a turbine. On overcast days, operators aim the mirrors skyward; sudden sun through clouds could heat the tower so quickly it could be destroyed.

(Continued from page 53) will always have to be supplemented by storage and other forms of power.

"We had a problem last night—no more tower," said Valerio Fernández, director of the plant, which is owned by Abengoa Solar, as he met me at the gate. He laughed as we looked up at PS10, its head lost in the clouds. On a normal day, the power focused on the tower could reach four megawatts per square meter—far more than can be safely utilized. PS10's operators have to limit the flux to avoid melting the receiver.

Power towers are a different version of solar thermal, another way to use sunlight to make steam. Although parabolic troughs are well proven for large, flat areas, power towers can be fit to hilly terrain, the mirrors individually aligned to converge on the elevated boiler. Because a tower heats steam to higher temperatures, it is potentially more efficient.

With the solar industry still in its infancy, however, Abengoa Solar is hedging its bets. Not far from the power towers, cranes were assembling rows of parabolic troughs. Behind PS10 stretched a field of advanced photovoltaics that track the sun on two axes—north-south as well as east-west—to ensure optimal exposure throughout the year. Each panel was fitted with mirrors or Fresnel lenses to intensify the light. "Taking profit from every one of the rays of sun—that's our goal," Fernández said.

Back home in the United States I read a magazine article challenging the country to move faster in harnessing the sun: "Every hour, it floods the earth with a deluge of thermal energy equal to 21 billion tons of coal," the writer had calculated. "The enormous output of solar energy is almost impossible to conceive." Illustrated with a drawing of a futuristic solar plant with enormous steam-generating mirrors, the

At a time of economic calamity, the New Deal transformed the nation's energy landscape. Seven decades later we still reap the benefits every time we flip a switch.

article was entitled "Why Don't We Have... Sun Power?" It was dated September 1953.

This time we might just make it. Last February, BrightSource Energy signed contracts with Southern California Edison for a series of power towers in southwestern deserts that could eventually provide 1.3 gigawatts of power, equal to a large coal-fired plant. Meanwhile, Pacific Gas and Electric has commissioned more than 1.8 gigawatts of parabolic troughs, photovoltaics, and BrightSource power towers. Environmentalists are already preparing to fight some of these projects; they would all cover large swaths of desert, and some might use a lot of scarce water for cooling. Like any form of power generation, solar has its trade-offs.

And it still has a long way to go. While I was in Nevada, I drove out to Hoover Dam—an early mass producer of renewable electricity—and joined a tour descending deep inside. At the bottom the torrent of Colorado River water falling from Lake Mead was spinning two parallel rows of giant turbines. Just one turbine puts out 130 megawatts, twice the power of Nevada Solar One.

But Hoover Dam left me feeling hopeful. Back on top, as I read the tarnished brass plaques and admired the art deco architecture, I thought about how this country had met the challenges of the Great Depression of the 1930s. The New Deal, as that earlier stimulus package was called, included not only Hoover but also the Tennessee Valley Authority, which brought hydroelectric power to the Southeast, and the Rural Electrification Administration, which strung power lines into the heartland. At a time of economic calamity, the nation's energy landscape was transformed. Seven decades later we still reap the benefits every time we flip a switch.

Mirrors concentrate solar heat to create turbine-spinning steam.

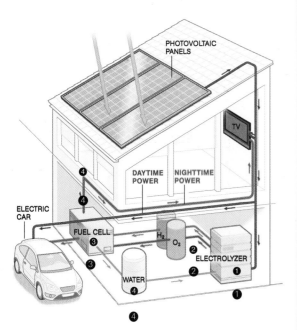

PHOTOVOLTAIC PANELS

TV

DAYTIME POWER | NIGHTTIME POWER

④

④

ELECTRIC CAR

FUEL CELL ③

③

WATER ④

H₂ O₂ ②

② ②

ELECTROLYZER ①

①

④

❶ Excess daytime electricity from solar panels goes to an electrolyzer.

❷ Aided by the new catalyst, the solar electricity splits water into hydrogen and oxygen, which are stored.

❸ When it's too dark for solar power, the stored elements are recombined in a fuel cell, generating electricity.

❹ Power flows to household appliances and recharges electric-car batteries. The only byproduct from the fuel cell—water—is recycled.

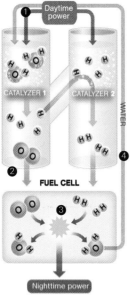

ELECTROLYZER

❶ Daytime power

CATALYZER 1   CATALYZER 2

WATER

❷

❹

FUEL CELL

❸

Nighttime power

An illustration of how solar homes are powered after dark.

Graphic: 5W INFOGRAPHICS

SOLAR 24/7 Power after dark remains a challenge for the solar home. But a cheap, self-renewing catalyst discovered by MIT researcher Daniel Nocera might allow water to act as a storage medium, keeping the lights on at night and even refueling an electric or hydrogen car. "Your house becomes a power plant," Nocera says. "It becomes a gas station."

Career Investigation

With the world's fossil fuel prices endlessly climbing skyward and the limited efficiencies of geothermal and wind energies, engineers are reinvesting their efforts to maximize the efficiencies of solar energy. The possibilities are endless as are the potential career opportunities.

Listed here are several possible careers to investigate. You may also be able to find a career path not listed. Choose three possible career paths and investigate what you would need to do to be ready to fill one of those positions.

Among the items to look for:

- Education—Does this require a college degree? What would you major in? Should you have a minor?

- Working conditions—What is the day to day job like? Will you have to be out in the field all day, or is it a desk job? Is the work strenuous? Are you working on a drilling rig at sea, away from your family for 3–6 months at a time? Is the job "hazardous"?

- Pay scale—What is the average "starting pay" for the position you are seeking? Don't be fooled by looking at "average pay" which may include those who have been working for 20+ years.

- Is the job located in the states, or is there a possibility to travel to other parts of the world?

- Are there any other special requirements?

Architect
Architectural Technician or Technologist
Civil Engineer
Civil Engineering Technician
Construction Manager
Electrician
Electricity Distribution Worker
Electronics Engineer
Engineering Construction Technician
Geoscientist
Geotechnician
Mechanical Engineering Technician
Physicist
Plumber
Project Manager
Solar Designer/Engineer
Solar Installation Manager
Solar PV Installer
Solar Thermal Installer
Structural Engineer
Thermal Insulation Engineer
Thermodynamics Designer
Welder

Team Building Activity

You will be assigned to a team of four or five students and your mission is to build a solar water heater. Your team will research and design a simple solar collector that will heat a one gallon container of water (at room temperature, supplied by the instructor) using only the sun's energy. Your instructor will take into consideration how quickly the water heats, the maximum temperature reached, and how long the water retains the heat once the sun's energy is removed. All of the groups will display and conduct their trials at the same time so as to eliminate variables such as ambient temperature, clouds, and other variables. While the experiments are running, groups will present their design, explaining the materials used and why they chose that particular design.

Writing Assignment

After reading the article "Plugging into the Sun" and conducting your own research, write a brief paper discussing the history of solar energy, current trends in solar energy, and where the future of solar energy is headed.

In addition to your own ideas and thoughts, please discuss:

- Is the United States ahead or behind in solar energy research? Please explain why or why not.

- In your conclusion, be sure to state whether you think the United States should pursue solar energy or should we focus our efforts on other renewable energies.

Listed here are a few links to help you in getting started.

Links:

ACCIONA's Nevada Solar One
http://www.acciona-na.com/About-Us/Our-Projects/U-S-/Nevada-Solar-One

Nevada Solar One—Acciona Gets Federal Grant to Expand
http://www.basinandrangewatch.org/SolarOneNV.html

Solar Steam at Nevada Solar One
http://solarpaces2008.sandia.gov/SolarPACES%20PLENARIES/2%20WEDNESDAY%20INDUSTRY%20DAY%20SESSIONS/1%20PLEN%20CSP%20PLANTS%20TODAY/01%20Acciona%20Cohen%20SOLARPACES%202008.pdf

Solar Electric Power Association
http://www.solarelectricpower.org/case-studies/nevada-solar-one.aspx

SPI Inc.
http://www.solarpowerinc.net/

Solar Energy
http://www.solarenergy.com/

Solar Energy Industries Association
http://www.seia.org/

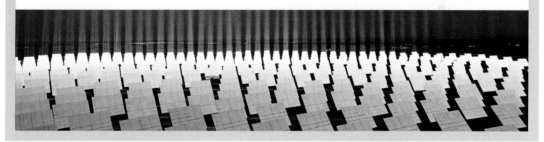

CAN SOLAR SAVE US?

By Chris Carroll

ANTICIPATION GUIDE

Purpose: To identify what you already know about solar energy, to direct and personalize your reading, and to provide a record of what new information you learned.

———◆———

Before you read "Can Solar Save Us?", examine each statement below and indicate whether you agree or disagree. Be prepared to discuss your reactions to the statements in groups.

- The sun can provide an endless supply of easily used energy forever.
- Solar energy is cheaper than using fossil fuels.
- President Obama wants Congress to require energy utilities to generate 25 percent of their energy from renewable (wind, water, solar, geothermal) sources by 2025.
- Italy, Germany, and Spain are leaders in utilizing solar energy.
- It is unrealistic to believe that the United States can switch to solar energy.
- Despite the cost of buying and installing solar panels, the utilization of solar energy is on the rise in the United States.
- With solar energy slow on the uptake in the United States, there is little reason to consider a career in solar power.
- There is tremendous potential for solar power in the United States.

PROBABLY. EVENTUALLY.
WITH LOTS OF GOVERNMENT HELP.

The sun is a utopian fuel: limitless, ubiquitous, and clean. Surely someday we'll give up on coal, oil, and gas—so hard on the climate, so unequally distributed worldwide—and go straight to the energy source that made fossil fuels. In a few sunny places where electric rates are high, like Italy and Hawaii, solar energy is already on the verge of being competitive. But in most places the sun remains by far the most expensive source of electric power—typically in the U.S. it costs several times more than natural gas or coal—which is why it still supplies only a fraction of a percent of our needs.

That won't change fast unless governments give solar a big boost. President Barack Obama campaigned with a pledge to institute a federal "renewable portfolio standard" requiring utilities to generate a quarter of their electricity from renewables by 2025. Yet even if Congress enacted that ambitious

In most places the sun remains by far the most expensive source of electric power.

law, coal would still dominate the nation's electricity portfolio two decades from now, and solar energy would probably remain a minor contributor. Cap-and-trade legislation that sets a price on carbon emissions would not be a magic bullet for solar either. Both mandates would likely lead utilities to favor the cheapest renewables, like wind. Solar would make a sizable contribution only after 2025, once the expansion of wind energy had plateaued.

Some advocates say we need to encourage solar more directly. European nations have done so with "feed-in tariffs," laws that require electric utilities to pay premiums to solar-power producers, be they commercial power plants or private homes that pump energy to the grid. Such tariffs have made Germany and Spain *(Continued on page 67)*

Adapted from "Can Solar Save Us" by Chris Carrol: National Geographic Magazine, September 2009.

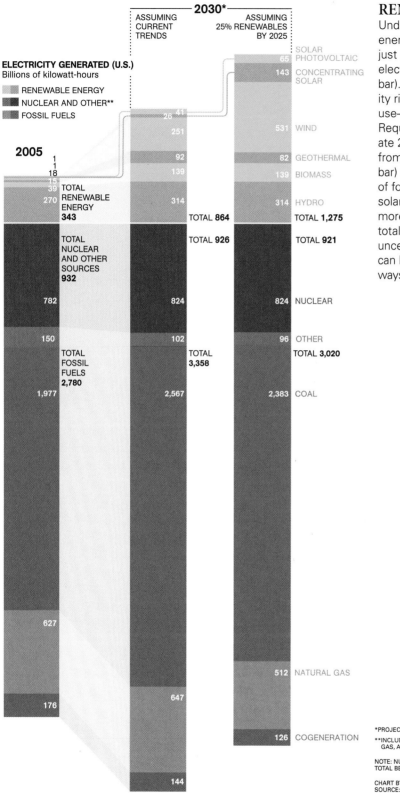

ELECTRICITY GENERATED (U.S.)
Billions of kilowatt-hours

- RENEWABLE ENERGY
- NUCLEAR AND OTHER**
- FOSSIL FUELS

2030*

ASSUMING CURRENT TRENDS

ASSUMING 25% RENEWABLES BY 2025

SOLAR PHOTOVOLTAIC 65

CONCENTRATING SOLAR 143

26 41

251 WIND 531

92 GEOTHERMAL 82

139 BIOMASS 139

2005

1
1
18
15
39
270

TOTAL RENEWABLE ENERGY **343**

314 HYDRO 314

TOTAL **864** TOTAL **1,275**

TOTAL NUCLEAR AND OTHER SOURCES **932**

TOTAL **926** TOTAL **921**

782 824 824 NUCLEAR

150 102 96 OTHER

TOTAL FOSSIL FUELS **2,780**

TOTAL **3,358** TOTAL **3,020**

1,977 2,567 2,383 COAL

627

512 NATURAL GAS

176 647

126 COGENERATION

144

RENEWABLE FUTURES

Under current policies, solar energy is projected to supply just over one percent of U.S. electricity by 2030 (middle bar). If demand for electricity rises, so will fossil fuel use—and carbon emissions. Requiring utilities to generate 25 percent of their power from renewable sources (right bar) would limit the growth of fossil fuels while pushing solar to 4 percent and wind to more than 10 percent of the total. Such forecasts are highly uncertain; policies and markets can both evolve in unforeseen ways.

*PROJECTIONS BASED ON 2005 DATA
**INCLUDES NONBIOGENIC MUNICIPAL WASTE, REFINERY GAS, AND OTHER SOURCES

NOTE: NUMBERS MAY NOT ADD UP TO TOTAL BECAUSE OF ROUNDING.

CHART BY 5W INFOGRAPHICS
SOURCE: UNION OF CONCERNED SCIENTISTS

(Continued from page 65) solar leaders, creating a market that has helped drive down prices. The billions of dollars of tax credits and loan guarantees in the Obama stimulus package may have a similar effect. Another option is for the federal government to invest directly in solar—for example, says Ken Zweibel of George Washington University, by funding the construction of giant solar plants in the desert Southwest, along with the high-efficiency transmission lines needed to carry the power nationwide. In Zweibel's version of the future, the sun would satisfy more than two-thirds of U.S. electricity needs by 2050, for an investment of about $400 billion. "Compared to what we just paid for the financial bailout, it's pocket change," he says.

Career Investigation

With the ever-escalating warnings of global warming and the rising carbon dioxide emissions many countries are moving to renewable energies. While the potential of geothermal and wind energies is somewhat limited, the career possibilities for solar energy are endless.

Listed here are several possible careers to investigate. You may also be able to find a career path not listed. Choose three possible career paths and investigate what you would need to do to be ready to fill one of those positions.

Among the items to look for:

- Education—Does this job require a college degree? What would you major in? Should you have a minor?

- Working conditions—What is the day to day job like? Will you have to be out in the field all day, or is it a desk job? Is the work strenuous? Are you working on a drilling rig at sea, away from your family for 3–6 months at a time? Is the job "hazardous"?

- Pay scale—What is the average "starting pay" for the position you are seeking? Don't be fooled by looking at "average pay" which may include those who have been working for 20+ years.

- Is the job located in the states, or is there a possibility to travel to other parts of the world?

- Are there any other special requirements?

Architect

Architectural Technician or Technologist

Civil Engineer

Civil Engineering Technician

Construction Manager

Electrician

Electricity Distribution Worker

Electronics Engineer

Engineering Construction Technician

Geoscientist

Geotechnician

Mechanical Engineering Technician

Physicist

Plumber

Project Manager

Solar Designer/Engineer

Solar Installation Manager

Solar PV Installer

Solar Thermal Installer

Structural Engineer

Thermal Insulation Engineer

Thermodynamics Designer

Welder

Team Building Activity

You will be assigned to a team of four or five students and will host a discussion panel to announce the conversion or construction of a new solar powered electric generating plant. The panel will consist of at least one each of an executive of the company that will operate the plant, an engineer that will design the plant, a member of the local/state government, and a member of the marketing firm hired to promote the company. At the end of the presentation, members of the general public (your classmates) will ask questions regarding the proposed construction/conversion.

Writing Assignment

After reading the article "Can Solar Save Us?" and conducting your own research, write a brief persuasive paper to your congressman/congresswoman to convince him or her that the United States should pursue solar energy and reduce our dependence on fossil fuels for energy.

In addition to your own ideas and thoughts, please discuss:

- What are other countries doing to promote solar energy?

- Should the government subsidize solar energy or build solar generating plants directly?

Listed here are a few links to help you in getting started.

Links:

Solar Energy – NY Times
http://topics.nytimes.com/top/news/business/energy-environment/solar-energy/index.html

Energy Story – California
http://www.energyquest.ca.gov/story/chapter15.html

American Solar Energy Society
http://ases.org/

National Renewable Energy Laboratory
http://www.nrel.gov/learning/re_solar.html

Florida Solar Energy Center
http://www.fsec.ucf.edu/en/

Science Daily
http://www.sciencedaily.com/news/matter_energy/solar_energy/

Texas Solar Energy Society
http://www.txses.org/solar/

THE 21ST CENTURY GRID

By Joel Achenbach

ANTICIPATION GUIDE

Purpose: To identify what you already know about the electrical infrastructure, to direct and personalize your reading, and to provide a record of what new information you learned.

———————◆———————

Before you read "The 21st Century Grid," examine each statement below and indicate whether you agree or disagree. Be prepared to discuss your reactions to the statements in groups.

- The electricity that comes to my house is generated by the local power company.
- There is a major storm in Chicago and I am 1,200 miles away so I don't have to worry about losing power.
- Our electrical grid is state of the art technology.
- Human error caused a massive blackout in 1965 leaving more than 30 million people without power for more than 12 hours.
- The electricity we use in our homes and businesses is sitting there waiting for us to flip a switch and use it.
- The first electrical transmission lines (from the Pearl Street Station in Brooklyn) to light Thomas Edison's light bulbs in Drexel, Morgan & Co. used DC (direct current) electricity (the same as what is used by batteries).
- Our electrical grid relies on an enormous amount of human interaction to function.
- There are 12 distinct power grids across the United States.
- Since the 1965 power failure, there has not been a massive power outage in the United States.
- Most of the electricity generated in the United States is hydroelectricity (produced from water flowing and spinning turbine generators).
- Our electrical grid is adequate to handle "green energies" (wind, solar, etc.).

Mirrors catch sunlight at an electric plant in southern Spain.
© Michael Melford/National Geographic Stock

From PlayStations to iMacs, the small electronic devices that appeal to so many have a big impact. Worldwide they account for about 15 percent of residential electricity consumption.

CAN WE FIX THE INFRASTRUCTURE
THAT POWERS
OUR LIVES?

We are creatures of the grid. We are embedded in it and empowered by it. The sun used to govern our lives, but now, thanks to the grid, darkness falls at our convenience. During the Depression, when power lines first electrified rural America, a farmer in Tennessee rose in church one Sunday and said—power companies love this story—"The greatest thing on earth is to have the love of God in your heart, and the next greatest thing is to have electricity in your house." He was talking about a few lightbulbs and maybe a radio. He had no idea.

Juice from the grid now penetrates every corner of our lives, and we pay no more attention to it than to the oxygen in the air. Until something goes wrong, that is, and we're suddenly in the dark, fumbling for flashlights and candles, worrying about the frozen food in what used to be called (in pre-grid days) the icebox. Or until the batteries run dry in our laptops or smart phones, and we find ourselves scouring the dusty corners of airports for an outlet, desperate for the magical power of electrons.

> **B**ut the first thing a smart grid will do, if we let it, is turn us into savvier consumers of electricity.

The grid is wondrous. And yet—in part because we've paid so little attention to it, engineers tell us—it's not the grid we need for the 21st century. It's too old. It's reliable but not reliable enough, especially in the United States, especially for our mushrooming population of finicky digital devices. Blackouts, brownouts, and other power outs cost Americans an estimated $80 billion a year. And at the same time that it needs to become more reliable, the grid needs dramatic upgrading to handle a different kind of power, a greener kind. That means, among other things, more transmission lines to carry wind power and solar power from remote places to big cities.

Most important, the grid must get smarter. The precise definition of "smart" varies from one engineer to the next. The gist is that a smart grid would be more automated and more "self-healing," and so less prone to failure. It would be more tolerant of small-scale, variable power sources such as solar panels

Adapted from "The 21st Century Grid" by Joel Achenbach: National Geographic Magazine, July 2010.

and wind turbines, in part because it would even out fluctuations by storing energy—in the batteries of electric cars, according to one speculative vision of the future, or perhaps in giant caverns filled with compressed air.

But the first thing a smart grid will do, if we let it, is turn us into savvier consumers of electricity. We'll become aware of how much we're consuming and cut back, especially at moments of peak demand, when electricity costs most to produce. That will save us and the utilities money—and incidentally reduce pollution. In a way, we'll stop being mere passive consumers of electrons. In the 21st century we'll become active participants in the management of this vast and seemingly unknowable network that makes our civilization possible.

So maybe it's time we got to know it.

There are grids today on six continents, and someday Europe's may reach across the Mediterranean into Africa to carry solar power from the Sahara to Scandinavia. In Canada and the United States the grid carries a million megawatts across tens of millions of miles of wire. It has been called the world's biggest machine. The National Academy of Engineering calls it the greatest engineering achievement of the last century.

Thomas Edison, already famous for his lightbulb, organized the birth of the grid in 1881, digging up lower Manhattan to lay down copper wires inside brick tunnels. He constructed a power plant, the Pearl Street Station, in the shadow of the Brooklyn Bridge. On September 4, 1882, in the office of tycoon J. P. Morgan, Edison threw a switch. Hundreds of his bulbs lit up Drexel, Morgan & Co. and other offices nearby.

It took decades for electricity to expand from factories and mansions into the homes

> The electrical grid is basically 1960s technology. The Internet has passed it by. The meter on the side of your house is 1920s technology.

of the middle class. In 1920 electricity still accounted for less than 10 percent of the U.S. energy supply. But inexorably it infiltrated everyday life. Unlike coal, oil, or gas, electricity is clean at the point of use. There is no noise, except perhaps a faint hum, no odor, and no soot on the walls. When you switch on an electric lamp, you don't think of the huge, sprawling power plant that's generating the electricity (noisily, odoriferously, sootily) many miles away. Refrigerators replaced iceboxes, air conditioners replaced heat prostration, and in 1956 the electric can opener completed our emergence from the dark ages. Today about 40 percent of the energy we use goes into making electricity.

At first, utilities were local operations that ran the generating plant and the distribution. A patchwork of mini-grids formed across the United States. In time the utilities realized they could improve reliability and achieve economies of scale by linking their transmission networks. After the massive Northeast blackout of 1965, much of the control of the grid shifted to regional operators spanning many states. Yet today there is still no single grid in the United States there are three nearly independent ones—the Eastern, Western, and Texas Interconnections.

They function with antiquated technology. The parts of the grid you come into contact with are symptomatic. How does the power company measure your electricity usage? With a meter reader—a human being who goes to your home or business and reads the dials on a meter. How does the power company learn that you've lost power? When you call on the phone. In general, utilities don't have enough instantaneous information on the flow of current through their lines—many of those lines don't carry any data—and people and

A silhouette of a man climbing a high power electric line tower.
© Dawn Kish/National Geographic Stock

slow mechanical switches are too involved in controlling that flow.

"The electrical grid is still basically 1960s technology," says physicist Phillip F. Schewe, author of *The Grid*. "The Internet has passed it by. The meter on the side of your house is 1920s technology." Sometimes that quaintness becomes a problem. On the grid these days, things can go bad very fast.

When you flip a light switch, the electricity that zips into the bulb was created just a fraction of a second earlier, many miles away. Where it was made, you can't know, because hundreds of power plants spread over many states are all pouring their output into the same communal grid. Electricity can't be stored on a large scale with today's technology; it has to be used instantly. At each instant there has to be a precise balance between generation and demand over the whole grid. In control rooms around the grid, engineers constantly monitor the flow of electricity, trying to keep voltage and frequency steady and to avoid surges that could damage both their customers' equipment and their own.

When I flip a switch at my house in Washington, D.C., I'm dipping into a giant pool of electricity called the PJM Interconnection. PJM is one of several regional operators that make up the Eastern grid; it covers the District of Columbia and 13 states, from the Mississippi River east to New Jersey and all the way down to the Outer Banks of North Carolina. It's an electricity market that keeps supply and demand almost perfectly matched—every day, every minute, every fraction of a second—among hundreds of producers and distributors and 51 million people, via 56,350 miles of high-voltage transmission lines.

One of PJM's new control centers is an hour north of Philadelphia. Last February I went to visit it with Ray E. Dotter, a company spokesman. Along the way Dotter identified the power lines we passed under. There was a pair of 500-kilovolt lines linking the Limerick nuclear plant with the Whitpain substation. Then a 230-kilovolt line. Then another. Burying the ungainly lines is prohibitively expensive except in dense cities. "There's a need to build new lines," Dotter said. "But no matter where you propose them, people don't want them."

Dotter pulled off the turnpike in the middle of nowhere. A communications tower poked above the treetops. We drove onto a compound surrounded by a security fence. Soon we were in the bunker, built by AT&T during the Cold War to withstand anything but a direct nuclear hit and recently purchased by PJM to serve as its new nerve center.

Computers take data from 65,000 points on the system, he explained. They track the thermal condition of the wires; too much power flowing through a line can overheat it, causing the line to expand and sag dangerously. PJM engineers try to keep the current alternating at a frequency of precisely 60 hertz. As demand increases, the frequency drops, and if it drops below 59.95 hertz, PJM sends a message to power plants asking for

> There's a need to build new lines, but no matter where you propose them, people don't want them.

more output. If the frequency increases above 60.05 hertz, they ask the plants to reduce output. It sounds simple, but keeping your balance on a tightrope might sound simple too until you try it. In the case of the grid, small events not under the control of the operators can quickly knock down the whole system.

Which brings us to August 14, 2003. Most of PJM's network escaped the disaster, which started near Cleveland. The day was hot; the air conditioners were humming. Shortly after 1 p.m EDT, grid operators at First Energy, the regional utility, called power plants to plead for more volts. At 1:36 p.m. on the shore of Lake Erie, a power station whose operator had just promised to "push it to my max max" responded by crashing. Electricity surged into northern Ohio from elsewhere to take up the slack.

At 3:05 a 345-kilovolt transmission line near the town of Walton Hills picked that moment to short out on a tree that hadn't been trimmed. That failure diverted electricity onto other lines, overloading and overheating them. One by one, like firecrackers, those lines sagged, touched trees, and short-circuited.

Grid operators have a term for this: "cascading failures." The First Energy operators couldn't see the cascade coming because an alarm system had also failed. At 4:06 a final line failure sent the cascade to the East Coast. With no place to park their electricity, 265 power plants shut down. The largest blackout in North American history descended on 50 million people in eight states and Ontario.

At the Consolidated Edison control center in lower Manhattan, operators remember that afternoon well. Normally the power load there dips gradually, minute by minute, as workers in the (Continued on page 82)

OUR SPRAWLING, EVOLVING GRID

A MAP OF THE EXISTING ELECTRICAL GRID (on the next page) may look at first glance like the interstate highway system, but the resemblance is superficial. Unlike the interstates, no one planned the grid. It consists of largely separate regional "interconnections" (below) that evolved from a patchwork of local utilities as they established links with their neighbors. Most of the grid is owned by those utilities; individual states regulate the construction of new transmission lines. Adding renewable energy to the grid and making it more reliable will require many new lines. That process is just beginning (see pages 78–81), and it too is happening without a nationwide plan.

QUEBEC
INTERCONNECTION

WESTERN
INTERCONNECTION

EASTERN
INTERCONNECTION

TEXAS
INTERCONNECTION

MARTIN GAMACHE & SAM PEPPLE, NGM STAFF

SOURCES: NORTH AMERICAN ELECTRIC RELIABILITY
CORPORATION; PLATTS, A DIVISION OF MCGRAW-HILL
COMPANIES (GENERATION AND TRANSMISSION
INFRASTRUCTURE); U.S. - CANADA POWER SYSTEM
OUTAGE TASK FORCE; U.S. DEPARTMENT OF ENERGY

California imports more electricity than any other state. Sources include hydroelectric plants in the Pacific Northwest and coal-burning ones in the desert Southwest. The Oregon-southern California link is the largest single transmission line in the U.S.

THE GRID TODAY

More than 150,000 miles of high-voltage transmission lines carry power from 5,400 generating plants owned by more than 3,000 utilities. Most of those lines carry alternating current (AC), but 1.9 percent of them carry direct current (DC), which loses less power over very long distances. The grid works 99.97 percent of the time—but power interruptions still cost the U.S. economy about $80 billion each

2003 BLACKOUT

Eight states and Ontario, Canada, (purple area) lost power in the 2003 blackout. It was a dramatic reminder of the vulnerability of the existing grid.

Blackout of 2003

Thunder Bay

Lake Superior

Duluth

Minneapolis

L. Huron

Lake Michigan

Milwaukee

Des Moines

Chicago

Detroit

L. Erie

Cleveland

Walton Hills

Quebec

Charlottetown

Halifax

Montreal

Ottawa

Toronto

L. Ontario

Niagara Falls

Buffalo

Boston

New York

Philadelphia

Pittsburgh

Cincinnati

St. Louis

Washington, D.C.

Norfolk

Raleigh

Nashville

Memphis

Atlanta

Jackson

New Orleans

Tampa

Miami

Power plants, 2009
(in megawatts)

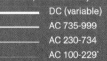

☐ Greater than 2,000

▫ 250 to 2,000

Transmission lines, 2009
(direct or alternating current, in kilovolts)

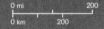

DC (variable)

AC 735-999

AC 230-734

AC 100-229*

* TRANSMISSION LINES BELOW 100 KV CAPACITY ARE NOT SHOWN.

MARTIN GAMACHE AND SAM PEPPLE, NGM STAFF

SOURCES: NORTH AMERICAN ELECTRIC RELIABILITY CORPORATION; PLATTS, A DIVISION OF MCGRAW-HILL COMPANIES (GENERATION AND TRANSMISSION INFRASTRUCTURE); U.S.-CANADA POWER SYSTEM OUTAGE TASK FORCE; U.S. DEPARTMENT OF ENERGY

0 mi   200

0 km   200

year. Moreover, our electricity is anything but clean. Most of it comes from burning fossil fuels, about half of it from coal. Hydroelectric, wind, and solar power account for less than 8 percent. The infrastructure perpetuates this: Texas currently has more wind-generation capacity than the grid can handle.

California's renewable energy law has led to a burst of wind and solar projects—as well as plans for high-voltage DC lines to bring renewable energy in from elsewhere. One idea: a 650-mile submarine cable to import hydroelectricity from Oregon.

Proposed new transmission lines, including the Tres Amigas project near Clovis, New Mexico, could help Texas deliver its abundant wind energy to far-flung cities in the East and West.

TEHACHAPI PROJECT

SOUTHWEST AREA NATL. INTEREST ELECTRIC TRANSMISSION CORRIDOR

NEW LINES ON THE GRID

Superimposing the grid of the future on the current one will not be cheap. Nearly $30 billion in new generation plants and high-voltage transmission lines are planned in the West alone. Both federal subsidies and state-set goals for renewable energy—30 percent in New York by 2015; 33 percent in California by 2020—are encouraging the construction of new transmission lines, which in some cases are also needed to improve the reliability of the grid. Reliability should also rise, along with energy efficiency, as the grid gets "smarter"—that is, as utilities increase their ability to monitor the flow of electricity from generator to consumer.

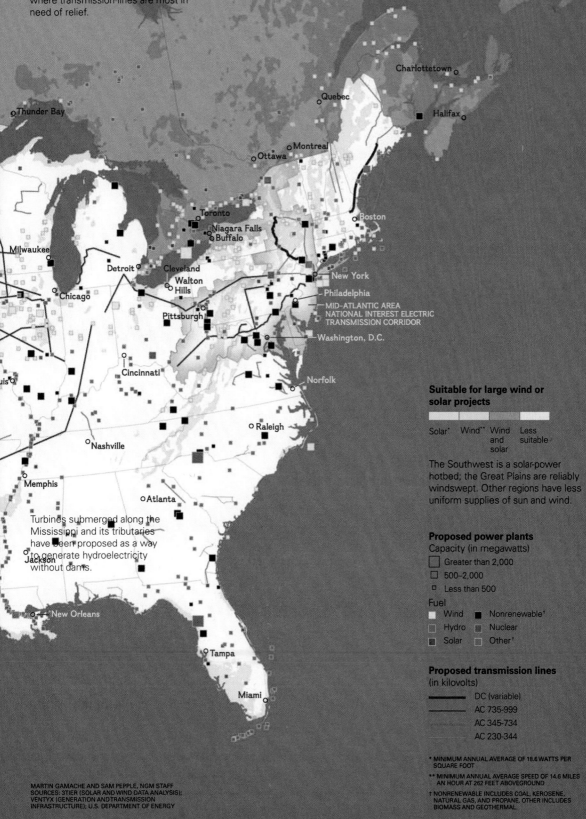

CONGESTED AREAS

The federal government has defined two "national corridors" (pink, below), in the mid-Atlantic and Southwest, where transmission lines are most in need of relief.

Thunder Bay

Charlottetown

Quebec

Halifax

Montreal

Ottawa

Boston

Toronto

Niagara Falls
Buffalo

Milwaukee

Detroit

Cleveland

Walton Hills

New York

Chicago

Pittsburgh

Philadelphia

MID-ATLANTIC AREA
NATIONAL INTEREST ELECTRIC
TRANSMISSION CORRIDOR

Cincinnati

Washington, D.C.

Norfolk

Raleigh

Nashville

Memphis

Atlanta

Turbines submerged along the Mississippi and its tributaries have been proposed as a way to generate hydroelectricity without dams.

Jackson

New Orleans

Tampa

Miami

Suitable for large wind or solar projects

| Solar* | Wind** | Wind and solar | Less suitable |

The Southwest is a solar-power hotbed; the Great Plains are reliably windswept. Other regions have less uniform supplies of sun and wind.

Proposed power plants
Capacity (in megawatts)

- Greater than 2,000
- 500–2,000
- Less than 500

Fuel

Wind	Nonrenewable†
Hydro	Nuclear
Solar	Other†

Proposed transmission lines
(in kilovolts)

▬▬▬	DC (variable)
———	AC 735-999
⋯⋯⋯	AC 345-734
------	AC 230-344

* MINIMUM ANNUAL AVERAGE OF 18.6 WATTS PER SQUARE FOOT

** MINIMUM ANNUAL AVERAGE SPEED OF 14.6 MILES AN HOUR AT 262 FEET ABOVEGROUND

† NONRENEWABLE INCLUDES COAL, KEROSENE, NATURAL GAS, AND PROPANE. OTHER INCLUDES BIOMASS AND GEOTHERMAL.

MARTIN GAMACHE AND SAM PEPPLE, NGM STAFF
SOURCES: 3TIER (SOLAR AND WIND DATA ANALYSIS);
VENTYX (GENERATION AND TRANSMISSION
INFRASTRUCTURE); U.S. DEPARTMENT OF ENERGY

(Continued from page 76) city turn off their lights and computers and head home. Instead, at 4:13 p.m. lights went out in the control room itself. The operators thought: 9/11. Then the phone rang, and it was the New York Stock Exchange. "What's going on?" someone asked. The operators knew at once that the outage was citywide.

There was no stock trading then, no banking, and no manufacturing; restaurants closed, workers were idled, and everyone just sat on the stoops of their apartment buildings. It took a day and a half to get power back, one feeder and substation at a time. The blackout cost six billion dollars. It also alarmed Pentagon and Homeland Security officials. They fear the grid is indeed vulnerable to terrorist attack, not just to untrimmed trees.

The blackout and global warming have provided a strong impetus for grid reform. The federal government is spending money on the grid—the economic-stimulus package allocated $4.5 billion to smart grid projects and another six billion dollars or so to new transmission lines. Nearly all the major utilities have smart grid efforts of their own.

A smarter grid would help prevent blackouts in two ways. Faster, more detailed feedback on the status of the grid would help operators stay ahead of a failure cascade. Supply and demand would also be easier to balance, because controllers would be able to tinker with both. "The way we designed and built the power system over the last hundred years—basically the way Edison and Westinghouse designed it—we create the supply side," says Steve Hauser of the U.S. Department of Energy's National Renewable Energy Laboratory (NREL) near Boulder, Colorado. "We do very little to control demand."

Working with the NREL, Xcel Energy has brought smart grid technology to Boulder. The first step is the installation of smart meters that transmit data over fiber-optic cable (it could also be done wirelessly) to the power company. Those meters allow consumers to see what electricity really costs at different times of day; it costs more to generate during times of peak load, because the utilities have to crank up auxiliary generators that aren't as efficient as the huge ones they run 24/7.

When consumers are given a price difference, they can choose to use less of the expensive electricity and more of the cheap kind. They can run clothes dryers and dishwashers at night, for instance. The next step is to let grid operators choose. Instead of only increasing electricity supply to meet demand, the operators could also reduce demand. On sweltering summer days the smart grid could automatically turn up thermostats and refrigerators a bit—with the prior agreement of the homeowners of course.

"Demand management" saves energy, but it could also help the grid handle renewable energy sources. One of the biggest problems with renewables like solar and wind power is that they're intermittent. They're not always available when demand peaks. Reducing the peak alleviates that problem. You can even imagine programming smart appliances to operate only when solar or wind power is available.

Some countries, such as Italy and Sweden, are ahead of the United States in upgrading their electrical intelligence. The Boulder project went online earlier this year, but only about 10 percent of U.S. customers have even the most primitive of smart meters, Hauser estimates.

"It's expensive," he says. "Utilities are used to spending 40 bucks on an old mechanical meter that's got spinning dials. A smart meter with a software chip, plus the wireless communication, might cost $200—five times as much. For utilities, that's huge." The Boulder project has cost *(Continued on page 84)*

A GRID THAT WORKS BOTH WAYS

A smart grid will change how the average home-owner thinks about electricity by constantly sharing usage data with the power company, which juggles supply (and pricing) accordingly. Consumers are rewarded for not hogging energy at times of peak demand; utilities benefit because power usage is more predictable and they learn immediately of any outages. Other improvements will make it easier to incorporate intermittent renewable energy sources such as wind and solar.

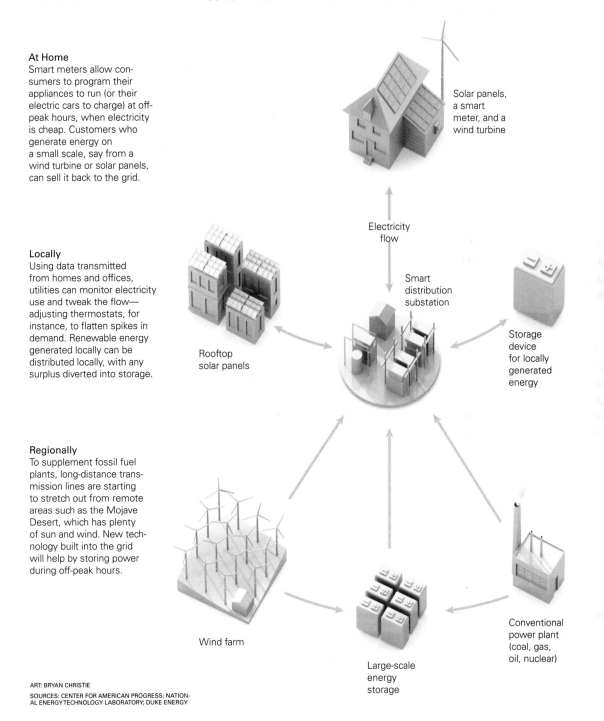

At Home
Smart meters allow con-sumers to program their appliances to run (or their electric cars to charge) at off-peak hours, when electricity is cheap. Customers who generate energy on a small scale, say from a wind turbine or solar panels, can sell it back to the grid.

Locally
Using data transmitted from homes and offices, utilities can monitor electricity use and tweak the flow—adjusting thermostats, for instance, to flatten spikes in demand. Renewable energy generated locally can be distributed locally, with any surplus diverted into storage.

Regionally
To supplement fossil fuel plants, long-distance trans-mission lines are starting to stretch out from remote areas such as the Mojave Desert, which has plenty of sun and wind. New tech-nology built into the grid will help by storing power during off-peak hours.

Solar panels, a smart meter, and a wind turbine

Electricity flow

Smart distribution substation

Storage device for locally generated energy

Rooftop solar panels

Wind farm

Large-scale energy storage

Conventional power plant (coal, gas, oil, nuclear)

ART: BRYAN CHRISTIE
SOURCES: CENTER FOR AMERICAN PROGRESS; NATION-AL ENERGY TECHNOLOGY LABORATORY; DUKE ENERGY

Stephen Heckeroth and his partner Gary drive one of their electric cars—a replica of James Dean's Porsche Spider. Heckeroth's primary goal is to do high profile projects that break our dependency on fossil fuels.
© Michael Nichols/National Geographic Stock

(Continued from page 82) Xcel Energy nearly three times what it expected. Earlier this year the utility raised rates to try to recoup some of those costs.

Although everyone acknowledges the need for a better, smarter, cleaner grid, the paramount goal of the utility industry continues to be cheap electricity. In the United States about half of it comes from burning coal. Coal-powered generators produce a third of the mercury emissions in the United States,

a third of our smog, two-thirds of our sulfur dioxide, and nearly a third of our planet-warming carbon dioxide—around 2.5 billion metric tons a year, by the most recent estimate.

Not counting hydroelectric plants, only about 3 percent of U.S. electricity comes from renewable energy. The main reason is that coal-fired electricity costs a few cents a kilowatt-hour, and renewables cost substantially more. Generally they're competitive only with the help of government regulations or tax incentives. Utility executives are a conservative bunch. Their job is to keep the lights on. Radical change makes them nervous; things they can't control, such as government policies, make them nervous. "They tend to like stable environments," says Ted Craver, head of Edison International, a utility conglomerate, "because they tend to make very large capital investments and eat that cooking for 30 or 40 or 50 years."

So windmills worry them. A utility executive might look at one and think: What if the wind doesn't blow? Or look at solar panels and think: What if it gets cloudy? A smart grid alone can't solve the intermittence problem. The ultimate solution is finding ways to store large amounts of electricity for a rainy, windless day.

Actually the United States can already store around 2 percent of its summer power output—and Europe even more—behind hydroelectric dams. At night, when electricity is cheaper, some utilities use it to pump water back uphill into their reservoirs, essentially storing electricity for the next day. A small power plant in Alabama does something similar; it pumps air into an underground cavern at night, compressing it to more than a thousand pounds per square inch. During the day the compressed air comes rushing out and spins a turbine. In the past year the Department of Energy has awarded stimulus money to several utilities for compressed-air projects. One project in Iowa would use wind energy to compress the air.

Another way to store electricity, of course, is in batteries. For the moment, it makes sense on a large scale only in extreme situations. For example, the remote city of Fairbanks, Alaska, relies on a huge nickel-cadmium, emergency backup battery. It's the size of a football field.

Although everyone acknowledges the need for a better, smarter, cleaner grid, the paramount goal of the utility industry continues to be cheap electric.

Lithium-ion batteries have more long-term potential—especially the ones in electric or plug-in-hybrid cars. PJM is already paying researchers at the University of Delaware $200 a month to store juice in three electric Toyotas as a test of the idea. The cars draw energy from the grid when they're charging, but when PJM needs electricity to keep its frequency stable, the cars are plugged in to give some back. Many thousands of cars, the researchers say, could someday function as a kind of collective battery for the entire grid. They would draw electricity when wind and solar plants are generating, and then feed some back when the wind dies down or night falls or the sun goes behind clouds. The Boulder smart grid is designed to allow such two-way flow.

To accommodate green energy, the grid needs not only more storage but more high-voltage power lines. There aren't enough running to the places where it's easy to generate the energy. To connect wind farms in Kern County with the Los Angeles area, Southern California Edison, a subsidiary of Edison International, is building 250 miles of them, known as the Tehachapi Renewable Transmission Project. A California law requires utilities to generate at least 20 percent of their electricity from renewable sources as of this year.

Green energy would also get a boost if there were more and bigger connections between the three quasi-independent grids in the United States. West Texas is a Saudi Arabia of wind, but the Texas Interconnection by itself can't handle all that energy. A proposed project called the Tres Amigas Superstation would allow Texas wind—and Arizona sun—to supply Chicago or Los Angeles. Near Clovis, (Continued on page 89)

ENERGY STORAGE

Because wind and solar power are intermittent, the quest is on to find ways to store energy for round-the-clock distribution—fueled in part by $185 million in stimulus money. Here are four promising technologies.

Compressed air, stored underground
An Iowa wind farm plans to pump air into sandstone formations when there's wind. Later the air can be released to make electricity.

Sodium-sulfur batteries
Pioneered in Japan, this technology can store a large amount of energy in a small space, as lithium-ion batteries do for electric cars.

Pumped-storage hydropower
Water is pumped uphill into a reservoir when electricity demand is low and released again to turn a turbine when demand is high.

Flywheel storage
In one town in Australia, electricity from wind turbines powers flywheels; their spinning motion is used to regenerate electricity when it's needed.

ART: BRYAN CHRISTIE

SOURCES: U.S. DEPARTMENT OF ENERGY; ENERGY STORAGE ASSOCIATION; ESKOM

Turbines funnel wind from the Columbia River Gorge.

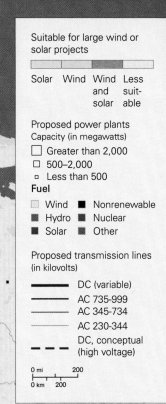

EUROPE'S GRID OF THE FUTURE

Last November 8, more than half of Spain's electricity was wind generated, and it even exported some to its neighbors—a hopeful sign, but one that lasted only a few hours. Europe's goal of getting 20 percent of its energy from renewables by 2020 will require "more of a revolution than an evolution," says Daniel Dobbeni of ENTSO-E, a grid operators association. Nine northern European countries agreed last year to link their grids by building transmission lines under the North Sea. A more futuristic vision: new lines under the Mediterranean to tap solar power from the Sahara.

NGM MAPS. SOURCES: 3TIER (SOLAR AND WIND DATA ANALYSIS); PLATTS, A DIVISION OF MCGRAW-HILL COMPANIES (PROPOSED GENERATION AND TRANSMISSION INFRA-STRUCTURE); DESERTEC; SUPERSMART GRID (DC CONCEPTUAL NETWORK)

Workers at an electric substation.
© Skip Brown/National Geographic Stock

(Continued from page 85) New Mexico, where the three interconnections already nearly touch, they would be joined together by a loop of five-gigawatt-capacity superconducting cable. The three grids would become, in effect, one single grid, national and almost rational.

The grid is a kind of parallel world that props up our familiar one but doesn't map onto it perfectly. It's a human construction that has grown organically, like a city or a government—what technical people call a kludge. A kludge is an awkward, inelegant contraption that somehow works. The U.S. grid works well by most measures, most of the time; electricity is abundant and cheap.

It's just that our measures have changed, and so the grid must too. The power industry, says Ted Craver of Edison International, faces "more change in the next ten years than we've seen in the last hundred." But at least now the rest of us are starting to pay attention.

Career Investigation

You come home from a long day at work and are about to prepare dinner and sit down to relax in front of the television. All of a sudden the power shuts off. You think little of it and are sure it will come back on momentarily. Even though it is hot outside, you leave the windows closed to retain the coolness from your air conditioner. After an hour you become concerned because the power is not yet back on. On your smart phone you access the local news and the station is reporting a major power outage over the entire western half of the United States. Furthermore, the news station reports that the power will be off for many hours to come. What about the food in the refrigerator, the air conditioning, the hot water heater, and the television? How will you cook and, most importantly, how will you charge your cell phone?

With our antiquated electrical system grid, there is great potential for new careers in designing, building, and perfecting new technologies to keep the electricity flowing to our homes.

Listed here are several possible careers to investigate. You may also be able to find a career path not listed. Choose three possible career paths and investigate what you would need to do to be ready to fill one of those positions.

Among the items to look for:

- Education—Does this job require a college degree? What would you major in? Should you have a minor?

- Working conditions—What is the day to day job like? Will you have to be out in the field all day, or is it a desk job? Is the work strenuous? Are you working on a drilling rig at sea, away from your family for 3–6 months at a time? Is the job "hazardous"?

- Pay scale—What is the average "starting pay" for the position you are seeking? Don't be fooled by looking at "average pay" which may include those who have been working for 20+ years.

- Is the job located in the states, or is there a possibility to travel to other parts of the world?

- Are there any other special requirements?

Architect
Architectural Technician or Technologist
Civil Engineer
Civil Engineering Technician
Construction Manager
Electrician
Electricity Distribution Worker
Electronics Engineer
Engineering Construction Technician
Geoscientist
Geotechnician
Lineman
Mechanical Engineering Technician
Physicist
Plumber
Project Manager
Solar Designer/Engineer
Solar Installation Manager
Solar PV Installer

Solar Thermal Installer
Steel Erector
Structural Engineer
Thermal Insulation Engineer
Thermodynamics Designer
Welder

Team Building Activity

You will be assigned to a team of four or five students to create and build an educational game designed around the information you researched for the "21st Century Grid" assignments. The board game can be designed after currently available board games (*Monopoly, Life,* etc.) or even a card-based game where players must collect sets of power transmission lines, electrical generation stations, fuel (coal, gas, hydroelectric, wind, solar, etc.) for the power generators, and so on. The game should be designed for rewards for correct choices (landing on the right spaces or collecting the correct sets of cards) and penalties for incorrect choices.

Writing Assignment

After reading the article "The 21st Century Grid" and conducting your own research, write a brief paper discussing the history of the "grid," the benefits of the "grid,"

and the problems we are currently facing with the "grid."

In addition to your own ideas and thoughts, please discuss:

- The need for a "smart grid."
- How the United States compares to other countries with respect to the electrical grid.
- The impact renewable energy sources will (or does) have on the grid.
- How your local utility company fits into the grid.
- Whether your local utility company plans on moving toward a "smart grid."

Listed here are a few links to help you in getting started.

Links:

Visual Power Grid of the United States
http://www.npr.org/templates/story/story.php?storyId=110997398

Forbes Article on the Grid
http://www.forbes.com/sites/williampentland/2012/03/20/plotting-the-next-generation-u-s-power-grid/

Researcher Modernizes U.S. Power Grid
http://www.sciencedaily.com/releases/2010/03/100330161753.htm

ANTICIPATION GUIDE

Purpose: To identify what you already know about fresh water, to direct and personalize your reading, and to provide a record of what new information you learned.

———————◆———————

Before you read "The Big Idea: Get the Salt Out" examine each statement below and indicate whether you agree or disagree. Be prepared to discuss your reactions to the statements in groups.

- There is plenty of "fresh water" on the planet Earth.
- Some people are drinking water from the sea or brackish water.
- The use of fresh water has increased proportionately to the population growth.
- One solution to supplying more fresh water is to remove the salt from seawater.
- Removing the salt from seawater is an easy and cheap process.
- We can use the sun's energy to create clean water from brackish water or seawater.

THE BIG IDEA: GET THE SALT OUT

By Karen E. Lange

97.5%
of the water
on Earth is salty.
Around 1 percent
of that is brackish
groundwater.

2.5%
of the Earth's water
is fresh. About two-
thirds of that is
frozen; the rest is
liquid surface water
and groundwater.

Wholesale water cost
in Southern California
(per 1,000 gallons)

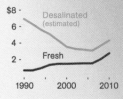

Better technology has
driven desalination
costs down—closer
to the price of fresh
water—though lately
they've risen again
with energy and
materials prices.

16 billion

gallons are produced
daily by the world's
14,450 desalination
plants. Persian Gulf
countries rely mostly
on seawater.

ART BY BRYAN CHRISTIE

SOURCES: TOM PANKRATZ, GLOBAL WATER
INTELLIGENCE; INTERNATIONAL DESALINA-
TION ASSOCIATION; MARK A. SHANNON,
UNIVERSITY OF ILLINOIS; ALEKSANDR NOY,
UNIVERSITY OF CALIFORNIA, MERCED

THERE'S NO SHORTAGE OF WATER ON THE BLUE PLANET—JUST A SHORTAGE OF FRESH WATER. NEW TECHNOLOGIES MAY OFFER BETTER WAYS TO...

As populations grow and agriculture and industry expand, fresh water—especially clean fresh water—is getting scarcer.

Three hundred million people now get their water from the sea or from brackish groundwater that is too salty to drink. That's double the number a decade ago. Desalination took off in the 1970s in the Middle East and has since spread to 150 countries. Within the next six years new desalination plants may add as much as 13 billion gallons a day to the global water supply, the equivalent of another Colorado River. The reason for the boom is simple: As populations grow and agriculture and industry expand, fresh water—especially clean fresh water—is getting scarcer. "The thing about water is, you gotta have it," says Tom Pankratz, editor of the Water Desalination Report, a trade publication. "Desalination is not a cheap way to get water, but sometimes it's the only way there is."

And it's much cheaper than it was two decades ago. The first desalination method—and still the most common, especially in oil-rich countries along the Persian Gulf—was brute-force distillation: Heat seawater until it turns to steam, leaving its salt behind, then condense it. The current state of the art, used, for example, at plants that opened recently in Tampa Bay, Florida, and Perth, Australia, is reverse osmosis, in which water is forced through a membrane that catches the salt. Pumping seawater to pressures of more than a thousand pounds per square inch takes less energy than boiling it—but it is still expensive.

Researchers are now working on at least three new technologies that could cut the energy required even further. The closest to commercialization, called forward osmosis, draws water through the porous membrane into a solution that contains even more salt than seawater, but a kind of salt that is easily evaporated. The other two approaches redesign the membrane itself—one by using

Adapted from "The Big Idea: Get the Salt Out" by Karen E. Lange: National Geographic Magazine, April 2010.

Three technologies promise to reduce the energy requirements of desalination by up to 30 percent. The race is on to see which will take the lead.

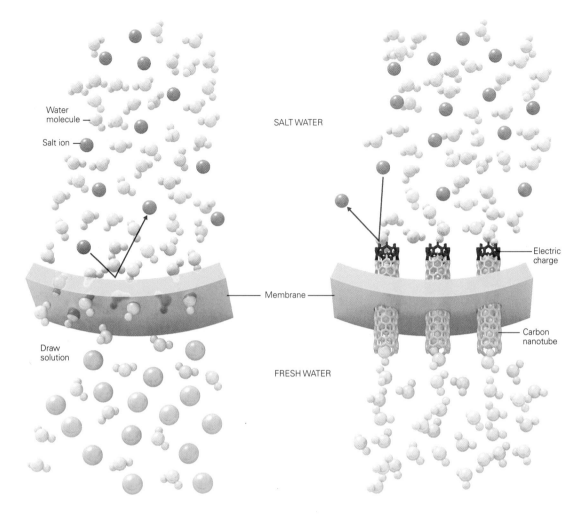

Water molecule

Salt ion

SALT WATER

Membrane

Electric charge

Draw solution

FRESH WATER

Carbon nanotube

FORWARD OSMOSIS
Water molecules migrate by natural osmosis, without energy input, into an even more concentrated "draw solution," whose special salt (green) is then evaporated away by low-grade heat.

On the market: 2010–2012

CARBON NANOTUBES
An electric charge at the nanotube mouth repels positively charged salt ions. The uncharged water molecules slip through with little friction, reducing pumping pressure.

On the market: 2013–2015

ART BY BRYAN CHRISTIE DESIGN SOURCES: TOM PANKRATZ, GLOBAL WATER INTELLIGENCE; INTERNATIONAL DESALINATION ASSOCIATION; MARK A. SHANNON, UNIVERSITY OF ILLINOIS; ALEKSANDR NOY, UNIVERSITY OF CALIFORNIA, MERCED

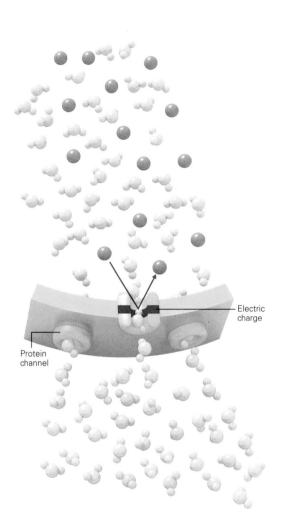

Electric charge

Protein channel

carbon nanotubes as the pores, the other by using the same proteins that usher water molecules through the membranes of living cells.

None of the three will be a solution for all the world's water woes. Desalination inevitably leaves behind a concentrated brine, which can harm the environment and even the water supply itself. Brine discharges are especially tricky to dispose of at inland desalination plants, and they're also raising the salinity in parts of the shallow Persian Gulf. The saltier the water gets, the more expensive it becomes to desalinate.

What's more, none of the new technologies seem simple and cheap enough to offer much hope to the world's poor, says geologist Farouk El-Baz of Boston University. He recently attended a desalination-industry conference looking for ways to bring fresh water to the war-torn Sudanese region of Darfur. "I asked the engineers, 'What if you are in a tiny village of 3,000, and the water is a hundred feet underground and laden with salt, and there is no electricity?'" El-Baz says. "Their mouths just dropped."

BIOMIMETICS
Water molecules pass through channels made of aquaporins, proteins that efficiently conduct water in and out of living cells. A positive charge near each channel's center repels salt.

On the market: 2013–2015

Career Investigation

On a hot, dry day, nothing slakes the thirst better than a cool glass of water. You go to the sink and turn on the faucet, only nothing comes out. The city's water supply is so severely stressed that water rationing is enforced and residential water is available only on even days of the week. While this may be unthinkable, many states are experiencing severe water shortages and are searching for solutions.

Listed here are several possible careers to investigate. You may also be able to find a career path not listed. Choose three possible career paths and investigate what you would need to do to be ready to fill one of those positions.

Among the items to look for:

- Education—Does this job require a specific college degree? What would you major in? Should you have a minor?

- Working conditions—What is the day to day job like? Will you have to be out in the field all day, or is it a desk job? Is the work strenuous? Are you working on a drilling rig at sea, away from your family for 3–6 months at a time? Is the job "hazardous"?

- Pay scale—What is the average "starting pay" for the position you are seeking? Don't be fooled by looking at "average pay" which may include those who have been working for 20+ years.

- Is the job located in the states, or is there a possibility to travel to other parts of the world?

- Are there any other special requirements?

Biologist
Biotechnologist
Civil Engineer
Desalination Plant Mechanic
Desalination Plant Operator
Desalination Plant Technician
Geoscientist
Geotechnician
Hydrologist
Land Surveyor
Landscape Manager
Landscape Scientist
Mechanical Engineer
Plumber
Research Engineer
Research Scientist
Structural Engineer
Town Planner
Town Planning Support Staff
Water Network Operative
Water Resource Manager
Water and Wastewater Treatment
    Operator

Team Building Activity

You will be assigned to a team of four or five students to host a panel that is composed of experts from around the world in the field of saltwater desalination. Your task is to announce the building of a new desalination plant.

The panel will present to an audience comprised of local townspeople and the news media. The team will use PowerPoint, posters, videos, and other supporting resources to solidify their points. Each team member should cover a specific category to solidify the position of the group either for or against additional construction.

After the presentation, the audience (representing members of the media and/or experts from the field) will be allowed to question the panel.

Writing Assignment

After reading the article "The Big Idea: Get the Salt Out" and conducting your own research, write a brief paper discussing the need for additional fresh water resources.

In addition to your own ideas and thoughts, please discuss:

- The impact that population has played on water consumption.

- The various methods of desalination.

- Other methods for supplying the world's population with fresh water.

Listed here are a few links to help you in getting started.

Links:

USGS—Earth's Water
http://ga.water.usgs.gov/edu/earthwherewater.html

Unwater.org
http://www.unwater.org/statistics_res.html

U-Mich.edu World's Fresh Water
http://www.globalchange.umich.edu/globalchange2/current/lectures/freshwater_supply/freshwater.html

The Facts About the Global Drinking Water Crisis
http://blueplanetnetwork.org/water/facts

The World's Water
http://www.worldwater.org/data.html

The Worldwide "Thirst" for Clean Drinking Water
http://www.npr.org/2011/04/11/135241362/the-worldwide-thirst-for-clean-drinking-water

Water Desalination Explained
http://geography.about.com/od/waterandice/a/Water-Desalination.htm

ANTICIPATION GUIDE

Purpose: To identify what you already know about deep water oil drilling, to direct and personalize your reading, and to provide a record of what new information you learned.

———————◆———————

Before you read "The Gulf of Oil: The Deep Dilemma," examine each statement below and indicate whether you agree or disagree. Be prepared to discuss your reactions to the statements in groups.

- Oil companies have been drilling for petroleum in deep water for many years.
- In the years between 1994 and 1997, the number of oil wells leased in the Gulf of Mexico in water half a mile deep or deeper increased 2,100 percent.
- As oil companies moved drilling operations to deeper water, the technology to drill deeper improved. The drilling companies also improved the methods of preventing blowouts and cleaning up spills.
- Drilling in the deep waters of the Gulf of Mexico can be highly productive, with a single well producing as much as 100,000 barrels a day.
- In 2009, the Deepwater Horizon drilling rig successfully drilled an oil-producing well in water that was 35,050 feet (approximately six miles) deep.
- The Deepwater Horizon drilling rig was responsible for the largest accidental oil spill in the world.
- BP (British Petroleum) which operated the Deepwater Horizon followed all of the regulations for deepwater drilling in February 2010.
- Deepwater drilling for oil carries a high risk of oil spills or oil well blowouts.
- After two and one half years the dangers from the oil spill (which was finally capped in July 2010) are finally over.
- Many people believe that the Deepwater Horizon event may result in more Americans becoming more concerned with our environment and switching to "green" energy alternatives.

THE GULF OF OIL: THE DEEP DILEMMA

By Joel K. Bourne, Jr.

Canals carved through Golden Meadow, La., and else-where hold pipelines that deliver oil and gas from offshore wells. This chopping up of the wetlands is one of many forces contributing to the decline of the Mississippi Delta.

A dead juvenile sea turtle lies marooned in oil in Barataria Bay, La. More than 500 sea turtles died in the spill area. As of August 2, eggs from 134 turtle nests had been moved to oil-free beaches, and 2,134 hatchlings released.

THE DEEP DILEMMA THE LARGEST U.S. OIL DISCOVERIES IN DECADES LIE IN THE DEPTHS OF THE GULF OF MEXICO—ONE OF

THE MOST DANGEROUS PLACES TO DRILL

ON THE PLANET.

Unflagging demand for oil propelled the industry into deep water but the blowout in the gulf forces the question: Is it worth the risk?

On a blistering June day in Houma, Louisiana, the local offices of BP—now the *Deepwater Horizon* Incident Command Center—were swarming with serious men and women in brightly colored vests. Top BP managers and their consultants wore white, the logistics team wore orange, federal and state environmental officials wore blue. Reporters wore purple vests so their handlers could keep track of them. On the walls of the largest "war room," huge video screens flashed spill maps and response-vessel locations. Now and then one screen showed a World Cup soccer match.

Mark Ploen, the silver-haired deputy incident commander, wore a white vest. A 30-year veteran of oil spill wars, Ploen, a consultant, has helped clean up disasters around the world, from Alaska to the Niger Delta. He now found himself surrounded by men he'd worked with on the *Exxon Valdez* spill in Alaska two decades earlier. "It's like a high school reunion," he quipped.

Fifty miles offshore, a mile underwater on the seafloor, BP's Macondo well was spewing something like an *Exxon Valdez* every four days. In late April an explosive blowout of the well had turned the *Deepwater Horizon*, one of the world's most advanced drill rigs, into a pile of charred and twisted metal at the bottom of the sea. The industry had acted as if such a catastrophe would never occur. So had its regulators. Nothing like it had happened in the Gulf of Mexico since 1979, when a Mexican well called Ixtoc I blew out in the shallow waters of the Bay of Campeche. Drilling technology had become so good since then, and the demand for oil so irresistible, that oil companies had sailed right off the continental shelf into ever deeper waters.

To many people in industry and government, spills from tankers like the *Exxon Valdez*

Adapted from "The Gulf of Oil: The Deep Dilemma" by Joel K. Bourne Jr.: National Geographic Magazine, October 2010.

seemed a much larger threat. The Minerals Management Service (MMS), the federal agency that regulated offshore drilling, had claimed that the chances of a blowout were less than one percent, and that even if one did happen, it wouldn't release much oil. Big spills had become a rarity, said Ploen. "Until this one."

In the Houma building, more than a thousand people were trying to organize a cleanup unlike any the world had seen. Tens of thousands more were outside, walking beaches in white Tyvek suits, scanning the waters from planes and helicopters, and fighting the expanding slick with skimmers, repurposed fishing boats, and a deluge of chemical dispersants. Around the spot Ploen called simply "the source," a small armada bobbed in a sea of oil. A deafening roar came from the drill ship *Discoverer Enterprise* as it flared off methane gas captured from the runaway well. Flames also shot from another rig, the *Q4000*, which was burning oil and gas collected from a separate line attached to the busted blowout preventer. Nearby, two shrimp boats pulling fire boom were burning oil skimmed from the surface, creating a curving wall of flame and a towering plume of greasy, black smoke. Billions of dollars had already been spent. But millions of barrels of light, sweet crude were still snaking toward the barrier islands, marshes, and beaches of the Gulf of Mexico.

The waters of the Gulf below a thousand feet are a relatively new frontier for oilmen—and one of the toughest places on the planet to drill. The seafloor falls off the gently sloping continental shelf into jumbled basin-and-range-like terrain, with deep canyons, ocean ridges, and active mud volcanoes 500 feet high. More than 2,000 barrels of oil a day seep from scattered natural vents. But the commercial deposits lie deeply buried, often beneath layers of shifting salt that are prone to undersea earth-

We have flipped design parameters around to the point that I got nervous. This has been [a] nightmare well.

quakes. Temperatures at the seafloor are near freezing, while the oil reservoirs can hit 400 degrees Fahrenheit; they're like hot, shaken soda bottles just waiting for someone to pop the top. Pockets of explosive methane gas and methane hydrates, frozen but unstable, lurk in the sediment, increasing the risk of a blowout.

For decades the exorbitant costs of drilling deep kept commercial rigs close to shore. But shrinking reserves, spiking oil prices, and spectacular offshore discoveries ignited a global rush into deep water. Recent finds in Brazil's Tupi and Guará fields could make that country one of the largest oil producers in the world. Similarly promising deepwater leases off Angola have excited bidding frenzies involving more than 20 companies.

In the Gulf of Mexico, the U.S. Congress encouraged companies to go deep as early as 1995. That year it passed a law forgiving royalties on deepwater oil fields leased between 1996 and 2000. A fleet of new rigs was soon punching holes all over the Gulf at a cost of up to a million dollars a day each. The number of leases sold in waters half a mile deep or more shot up from around 50 in 1994 to 1,100 in 1997.

As technology was taking drillers deeper, however, the methods for preventing blowouts and cleaning up spills did not keep pace. Since the early 2000s, reports from industry and academia warned of the increasing risk of deepwater blowouts, the fallibility of blowout preventers, and the difficulty of stopping a deepwater spill after it started—a special concern given that deepwater wells, because they're under such high pressure, can spout as much as 100,000 barrels a day.

The Minerals Management Service routinely downplayed such concerns. A 2007 agency study found that from 1992 to 2006, only 39 blowouts occurred during the drilling of more than 15,000 oil and gas wells in the Gulf. Few of them released much oil; only one resulted in a death. Most of the blowouts

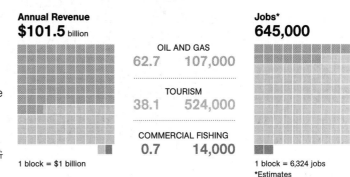

WORKING GULF

Oil dominates revenues from the Gulf, but the employment giant is tourism. Louisiana, regional leader in commercial fishing before the spill, normally harvests a third of the U.S. shrimp and oyster catch.

NGM ART,

SOURCES: EIA (OIL 2008); TOURISM DEPARTMENTS OF ALABAMA, LOUISIANA, MISSISSIPPI, AND TEXAS AND FLORIDA DEPARTMENT OF REVENUE (TOURISM 2009); NOAA (FISHING, 2008, DOCKSIDE VALUE); JOBS MOST RECENT AVAILABLE DATA FROM MULTIPLE SOURCES

Annual Revenue
$101.5 billion

OIL AND GAS
62.7   107,000

TOURISM
38.1   524,000

COMMERCIAL FISHING
0.7   14,000

1 block = $1 billion

Jobs*
645,000

1 block = 6,324 jobs
*Estimates

were stopped within a week, typically by pumping the wells full of heavy drilling mud or by shutting them down mechanically and diverting the gas bubble that had produced the dangerous "kick" in the first place.

Though blowouts were relatively rare, the MMS report did find a significant increase in the number associated with cementing, the process of pumping cement around the steel well casing (which surrounds the drill pipe) to fill the space between it and the wall of the borehole. In retrospect, that note of caution was ominous.

Some deepwater wells go in relatively easy. The Macondo well did not. BP hired Transocean, a Switzerland-based company, to drill the well. Transocean's first drill rig was knocked out of commission by Hurricane Ida after just a month. The *Deepwater Horizon* began its ill-fated effort in February 2010 and ran into problems almost from the start. In early March the drill pipe got stuck in the borehole, as did a tool sent down to find the stuck section; the drillers had to back out and drill around the obstruction. A BP email later released by Congress mentioned that the drillers were having "well-control" problems. Another email, from a consultant, stated, "We have flipped design parameters around to the point that I got nervous." A week before the explosion, a BP drilling engineer wrote, "This has been [a] nightmare well."

By April 20 the *Deepwater Horizon* was six weeks behind schedule, according to MMS documents, and the delay was costing BP more than half a million dollars a day. BP had chosen to drill the fastest possible way—using a well design known as a "long string" because it places strings of casing pipe between the oil reservoir and the wellhead. A long string generally has two barriers between the oil and the blowout preventer on the seafloor: a cement plug at the bottom of the well, and a metal seal, known as a lockdown sleeve, placed right at the wellhead. The lockdown sleeve had not been installed when the Macondo well blew out.

In addition, congressional investigators and industry experts contend that BP cut corners on its cement job. It failed to circulate heavy drilling mud outside the casing before cementing, a practice that helps the cement cure properly. It didn't put in enough centralizers—devices that ensure that the cement forms a complete seal around the casing. And it failed to run a test to see if the cement had bonded properly. Finally, just before the accident, BP replaced the heavy drilling mud in the well with much lighter seawater, as it prepared to finish and disconnect the rig from the well. BP declined to comment on these matters, citing the ongoing investigation.

All these decisions may have been perfectly legal, and they surely saved BP time and money—yet each increased the risk of a blowout. On the night of April 20, investigators suspect, a large gas bubble (Continued on page 108)

DRILLING FOR OFFSHORE OIL

Undersea oil provides an increasing amount of the global supply, as exploration heads ever deeper in search of new "plays." In 2020 wells more than 400 meters below the sea surface will likely provide 10 percent of the world's oil. But going deep poses technical challenges and safety risks.

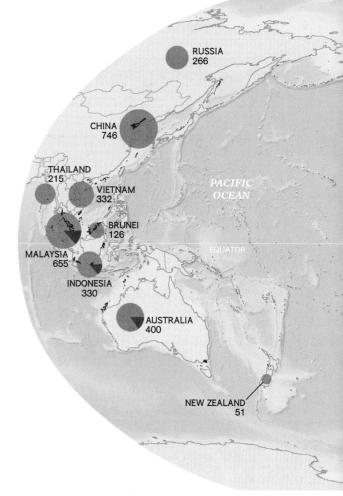

RUSSIA
266

CHINA
746

THAILAND
215

VIETNAM
332

BRUNEI
126

MALAYSIA
655

INDONESIA
330

AUSTRALIA
400

NEW ZEALAND
51

PACIFIC OCEAN

EQUATOR

Top ten offshore platform spills, 1969–2010
Millions of barrels

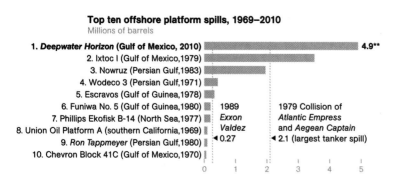

1. *Deepwater Horizon* (Gulf of Mexico, 2010)   4.9**
2. Ixtoc I (Gulf of Mexico, 1979)
3. Nowruz (Persian Gulf, 1983)
4. Wodeco 3 (Persian Gulf, 1971)
5. Escravos (Gulf of Guinea, 1978)
6. Funiwa No. 5 (Gulf of Guinea, 1980)
7. Phillips Ekofisk B-14 (North Sea, 1977)
8. Union Oil Platform A (southern California, 1969)
9. *Ron Tappmeyer* (Persian Gulf, 1980)
10. Chevron Block 41C (Gulf of Mexico, 1970)

1989
Exxon
Valdez
◄ 0.27

1979 Collision of
Atlantic Empress
and *Aegean Captain*
◄ 2.1 (largest tanker spill)

0   1   2   3   4   5

MARTIN GAMACHE, NGM STAFF,

SOURCES: PETER BURGHERR, PAUL SCHERRER INSTITUTE (PLATFORM SPILLS); FLOW RATE TASK GROUP (DEEPWATER ESTIMATES); IHS ENERGY (RESERVES); MICHAEL R. SMITH, DATAMONITOR, "GLOBAL OIL AND GAS ANALYZER" (2009 PRODUCTION)

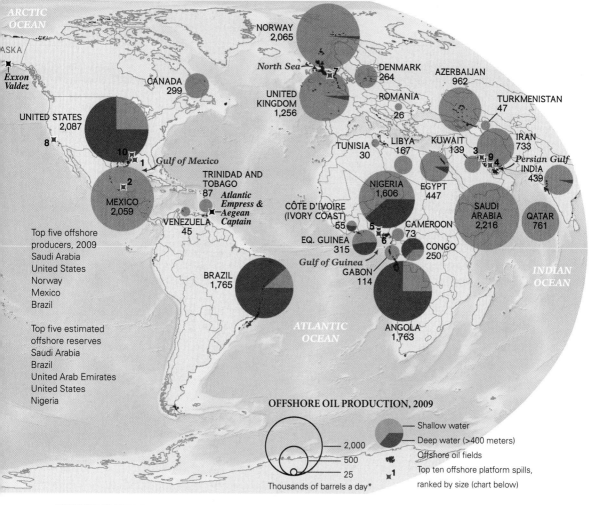

ARCTIC
OCEAN

ALASKA

Exxon
Valdez

UNITED STATES
2,087

8

CANADA
299

10
1   Gulf of Mexico

2

MEXICO
2,059

VENEZUELA
45

TRINIDAD AND
TOBAGO
87   Atlantic
Empress &
Aegean
Captain

NORWAY
2,065

North Sea   7

UNITED
KINGDOM
1,256

DENMARK
264

ROMANIA
26

TUNISIA
30

LIBYA
167

CÔTE D'IVOIRE
(IVORY COAST)
55

EQ. GUINEA
315

Gulf of Guinea
GABON
114

BRAZIL
1,765

NIGERIA
1,606

5

6

CAMEROON
73

EGYPT
447

AZERBAIJAN
962

TURKMENISTAN
47

KUWAIT
139   3

9

4

IRAN
733

Persian Gulf
INDIA
439

SAUDI
ARABIA
2,216

QATAR
761

CONGO
250

ANGOLA
1,763

ATLANTIC
OCEAN

INDIAN
OCEAN

Top five offshore
producers, 2009
Saudi Arabia
United States
Norway
Mexico
Brazil

Top five estimated
offshore reserves
Saudi Arabia
Brazil
United Arab Emirates
United States
Nigeria

OFFSHORE OIL PRODUCTION, 2009

2,000

500

25

Thousands of barrels a day*

Shallow water
Deep water (>400 meters)
Offshore oil fields
Top ten offshore platform spills,
1 ranked by size (chart below)

JUAN VELASCO, NGM STAFF, ART BY BRYAN CHRISTIE

SOURCES: RENAUD BOUROULLEC, COLORADO SCHOOL OF MINES AND PAUL WEIMER, UNIVERSITY OF COLORADO
(GEOLOGY AND BATHYMETRY); LOUSIANA DEPARTMENT OF NATURAL RESOURCES (SHALLOW-WATER WELLS);
MMS (DEEP AND ULTRADEEP WELLS, OIL FROM FEDERAL LEASES); ENERGY INFORMATION ADMINISTRATION, OR
BA (U.S. PRODUCTION)

*Only countries producing at least 25,000 barrels
a day are shown (one barrel = 42 U.S. gallons).
For legibility, some pie charts are shown inland
or outside their country boundaries.

**August 2, 2010, estimate; of this total,
800,000 barrels of oil were captured by BP
at the well.

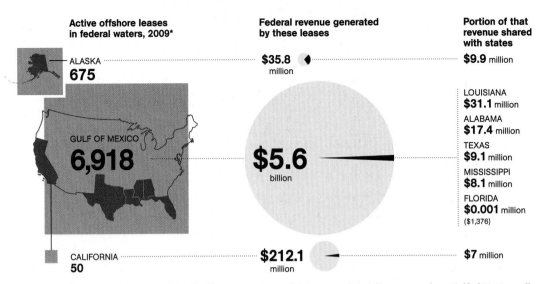

Active offshore leases in federal waters, 2009*	Federal revenue generated by these leases	Portion of that revenue shared with states
ALASKA **675**	**$35.8** million	**$9.9** million
GULF OF MEXICO **6,918**	**$5.6** billion	LOUISIANA **$31.1** million ALABAMA **$17.4** million TEXAS **$9.1** million MISSISSIPPI **$8.1** million FLORIDA **$0.001** million ($1,376)
CALIFORNIA **50**	**$212.1** million	**$7** million

Offshore leases poured $5.8 billion into federal coffers in 2009. Most of the money—$5.6 billion—came from Gulf of Mexico wells, which have helped drive offshore oil's contribution to domestic production to 35 percent, up from 12 percent in 1981. Leasing is expected to resume once drillers adopt new federal safety standards.

ALASKA AND CALIFORNIA DATA AS OF SEPTEMBER 30, 2009; GULF OF MEXICO DATA AS OF AUGUST 18, 2009

(Continued from page 105) somehow infiltrated the casing, perhaps through gaps in the cement, and shot straight up. The blowout preventer should have stopped that powerful kick at the seafloor; its heavy hydraulic rams were supposed to shear the drill pipe like a soda straw, blocking the upward surge and protecting the rig above. But that fail-safe device had itself been beset by leaks and maintenance problems. When a geyser of drilling mud erupted onto the rig, all attempts to activate the blowout preventer failed.

The way BP drilled the Macondo well surprised Magne Ognedal, director general of the Petroleum Safety Authority Norway (PSA). The Norwegians have drilled high-temperature, high-pressure wells on their shallow continental shelf for decades, he said in a telephone interview, and haven't had a catastrophic blowout since 1985. After that incident, the PSA and the industry instituted a number of best practices for drilling exploration wells. These include riserless drilling from stations on the seafloor, which prevents oil and gas from flowing directly to a rig; starting a well with a small pilot hole through the sediment, which makes it easier to handle gas kicks;

having a remote-controlled backup system for activating the blowout preventers; and most important, never allowing fewer than two barriers between the reservoir and the seafloor.

"The decisions [BP] made when they had indications that the well was not stable, the decision to have one long pipe, the decision to have only six centralizers instead of 21 to create the best possible cement job—some of these things were very surprising to us here," says Ognedal.

The roots of those decisions lie in BP's corporate history, says Robert Bea, a University of California, Berkeley expert in both technological disasters and offshore engineering. BP hired Bea in 2001 for advice on problems it faced after it took over the U.S. oil companies Amoco and ARCO. One problem, Bea says, was a loss of core competence: After the merger BP forced thousands of older, experienced oil field workers into early retirement. That decision, which made the company more dependent on contractors for engineering expertise, was a key ingredient in BP's "recipe for disaster," Bea says. Only a few of the 126 crew members on the *Deepwater Horizon* worked directly for BP.

The drilling operation itself was regulated by the MMS (which, in the wake of the accident, was reorganized and renamed the Bureau of Ocean Energy Management, Regulation, and Enforcement). In 2009 the MMS had been excoriated by the U.S. General Accounting Office for its lax oversight of offshore leases. That same year, under the new Obama Administration, the MMS rubber-stamped BP's initial drilling plan for the Macondo well. Using an MMS formula, BP calculated that the worst-case spill from the well would be 162,000 barrels a day—nearly three times the flow rate that actually occurred. In a separate spill-response plan for the whole Gulf, the company claimed that it could recover nearly 500,000 barrels a day using standard technology, so that even a worst-case spill would do minimal harm to the Gulf's fisheries and wildlife—including walruses, sea otters, and sea lions.

There are no walruses, sea otters, or sea lions in the Gulf. BP's plan also listed as an emergency responder a marine biologist who had been dead for years, and it gave the Web address of an entertainment site in Japan as an emergency source of spill-response equipment. The widely reported gaffes had appeared in other oil companies' spill-response plans as well. They had simply been cut and pasted from older plans prepared for the Arctic.

When the spill occurred, BP's response fell well short of its claims. Scientists on a federal task force said in early August that the blownout well had disgorged as much as 62,000 barrels a day at the outset—an enormous flow rate, but far below BP's worst-case scenario. Mark Ploen estimated in June that on a good day his response teams, using skimmers brought in from around the world, were picking up 15,000 barrels. Simply burning the oil, a practice that had been used with the *Exxon Valdez* spill, had proved more effective. BP's burn fleet of 23 vessels included local shrimp boats that worked in pairs, corralling surface oil with long fire boom and then igniting it with homemade napalm. In one "monster burn" the team incinerated 16,000 barrels of oil in just over three hours.

"Shrimpers are naturals at doing this," said Neré Mabile, science and technology adviser for the burn team in Houma. "They know how to pull nets. They're seeing that every barrel we burn is a barrel that doesn't get to shore, doesn't affect the environment, doesn't affect people. And where's the safest place to burn this stuff? The middle of the Gulf of Mexico."

In June the *Discoverer Enterprise* and the *Q4000* began collecting oil directly at the busted blowout preventer, and by mid-July they had ramped up to 25,000 barrels a day—still far less, even when the efforts of the skimmers and the burn team were added, than the nearly 500,000 barrels a day BP had claimed it could remove. At that point the company finally succeeded in placing a tight cap on the well, halting the gusher after 12 weeks.

In 1990, after the *Exxon Valdez* spill, Congress's Office of Technology Assessment analyzed spill-response technologies and found them lacking. "Even the best national response system will have inherent practical limitations that will hinder spill-response efforts for catastrophic events—sometimes to a major extent," wrote OTA's director, John H. Gibbons. "For that reason it is important to pay at least equal attention to preventive measures as to response systems…The proverbial ounce of prevention is worth many, many pounds of cure."

By early August BP seemed on the verge of plugging the Macondo well permanently with drilling mud and cement. The federal task force's estimate of the amount of oil released stood at 4.9 million barrels.

When oil falls to the bottom, into the mud of lagoon or a marsh, it can hang around for decades, degrading the environment.

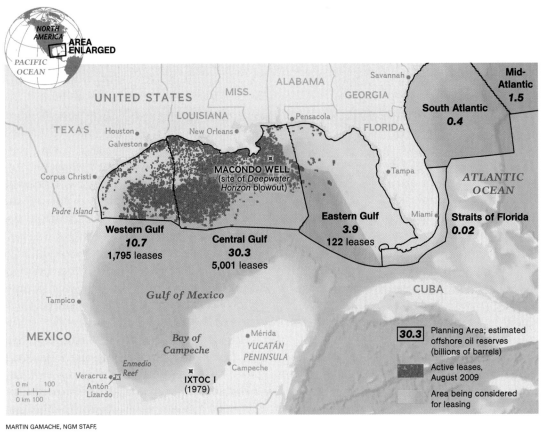

MARTIN GAMACHE, NGM STAFF,

SOURCES: BUREAU OF OCEAN ENERGY MANAGEMENT, REGULATION AND ENFORCEMENT, FORMERLY MINERALS MANAGEMENT SERVICE OR MMS (REVENUE AND LEASES); CONGRESSIONAL RESEARCH SERVICE (RESERVE ESTIMATES)

Government scientists estimated that BP had removed a quarter of the oil. Another quarter had evaporated or dissolved into scattered molecules. But a third quarter had been dispersed in the water as small droplets, which might still be toxic to some organisms. And the last quarter—around five times the amount released by the *Exxon Valdez*—remained as slicks or sheens on the water or tar balls on the beaches. The *Deepwater Horizon* spill had become the largest accidental spill into the ocean in history, larger even than the Ixtoc I blowout in Mexico's Bay of Campeche in 1979. It is surpassed only by the intentional 1991 gulf war spill in Kuwait.

The Ixtoc spill devastated local fisheries and economies. Wes Tunnell remembers it well. The tall, 65-year-old coral reef expert at Texas A&M University–Corpus Christi earned his doctorate studying the reefs around Veracruz in the early 1970s, and he

kept studying them for a decade after the spill coated them with oil. Tunnell wrote an early report on the impact there and on Padre Island in Texas. In early June, after the new disaster had once again raised the question of how long the impact of a spill can last, he returned to Enmedio Reef to see if any Ixtoc I oil remained. It took him three minutes of snorkeling to find some. "Well, that was easy," he said.

Tunnell stood in the clear, waist-deep water of the protected reef lagoon holding what appeared to be a three-inch-thick slab of sandy gray clay. When he broke it in two, it was jet black on the inside, with the texture and smell of an asphalt brownie. Here on the lagoon side, where the reef looked gray and dead, the Ixtoc tar mat was still partially buried in the sediments. But on the ocean side of the reef, where winds and waves and currents were stronger, no oil remained. The lesson for Louisiana and the other Gulf states is clear, Tunnell thinks. Where there is wave energy and oxygen, sunlight and the Gulf's abundant oil-eating bacteria break it down fairly quickly. When oil falls to the bottom and gets entrained in low-oxygen sediments like those in a lagoon—or in a marsh—it can hang around for decades, degrading the environment.

Fishermen in the nearby village of Antón Lizardo hadn't forgotten the spill either. "The Ixtoc spill about destroyed all the reefs," said Gustavo Mateos Moutiel, a powerful man, now in his 60s, who wore the trademark straw hat of the Veracruzano fishermen. "Octopus gone. Urchins gone. Oysters gone. Conch gone. Fish almost all gone. Our families were hungry. The petroleum on the beach was halfway up our knees." Though some species, such as Bay of Campeche shrimps, recovered within a few years, Moutiel, along with several other fishermen who had gathered on the beach, said it took 15 to 20 years for their catches to return to normal. By then two-thirds of the fishermen in the village had found other jobs.

Even in the turbulent, highly oxygenated waters of France's Breton coast, it took at least seven years after the 1978 Amoco Cadiz spill for local marine species and Brittany's famed oyster farms to fully recover, according to French biologist Philippe Bodin. An expert on marine copepods, Bodin studied the long-term effects of the spill from the grounded tanker. He believes the impact will be far worse in the generally calmer, lower-oxygen waters of the Gulf, particularly because of the heavy use of the dispersant Corexit 9500. BP has said the chemical is no more toxic than dish-washing liquid, but it was used extensively on the Amoco Cadiz spill, and Bodin found it to be more toxic to marine life than the oil itself. "The massive use of Corexit 9500 in the Gulf is catastrophic for the phytoplankton, zooplankton, and larvae," he says. "Moreover, currents will drive the dispersant and the oil plumes everywhere in the Gulf."

In May, scientists in the Gulf began tracking plumes of methane and oil droplets drifting up to 30 miles from the broken well, at depths of 3,000 to 4,000 feet. One of those scientists was University of Georgia biogeochemist Mandy Joye, who has spent years studying hydrocarbon vents and brine seeps in the deep Gulf. She found a plume the size of Manhattan, and its methane levels were the highest she had ever measured in the Gulf. As bacteria feast on spilled oil and methane, they deplete the water of oxygen; at one point Joye found oxygen levels dangerously low for life in a water layer 600 feet thick, at depths where fish usually live. Since waters in the deep Gulf mix very slowly, she said, such depleted zones could persist for decades.

BP was using old DC-3s set up like giant crop dusters to spray Corexit 9500 onto surface slicks. But for the world's first major deepwater spill, the company also got permission from the U.S. Environmental Protection Agency and the Coast Guard to pump hundreds of thousands of gallons of dispersant directly into the oil and gas spewing from the well, a mile beneath the surface. That helped create the deepwater plumes.

"The whole goal is to keep oil off the beaches, because that's what drives the economy," Joye said one day in June as she ran samples through her gas chromatograph aboard the R.V. F. G. Walton Smith. The little research ship was bobbing in an oily sheen a few miles from the busted well. "But now you've got all this material in the water column that no one is seeing and that you can't get rid of. If oil gets to the surface, about 40 percent evaporates. You can skim it, you can burn it, you can do something with it. But these tiny particles in the water column will persist for God knows how long."

Oceanographer Ian MacDonald at Florida State University worries not only about the plumes but also about the sheer volume of spilled oil. He believes it could have a major impact on the overall productivity of the Gulf—not just on pelicans and (Continued on page 116)

DRILLING DEEPER

As oil and gas reserves close to shore have been pumped dry, prospectors are plumbing a new frontier: the depths of the Gulf of Mexico. In 2009 Gulf oil production jumped 34 percent—largely from waters deeper than 5,000 feet. New technologies have made it possible to drill more than 35,000 feet down through water and rock.

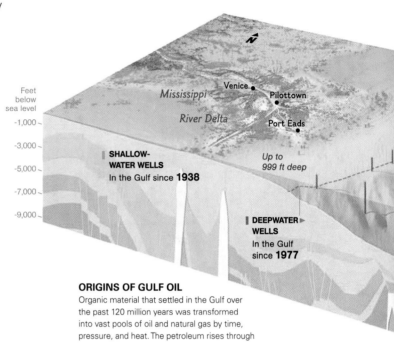

Feet below sea level
-1,000
-3,000
-5,000
-7,000
-9,000

Venice
Pilottown
Mississippi
River Delta
Port Eads

SHALLOW-WATER WELLS
In the Gulf since **1938**

Up to 999 ft deep

DEEPWATER WELLS
In the Gulf since **1977**

ORIGINS OF GULF OIL

Organic material that settled in the Gulf over the past 120 million years was transformed into vast pools of oil and natural gas by time, pressure, and heat. The petroleum rises through faults until it is trapped by salt structures, some more than a mile below the seafloor.

JUAN VELASCO, NGM STAFF. ART BY BRYAN CHRISTIE

SOURCES: RENAUD BOUROULLEC, COLORADO SCHOOL OF MINES, AND PAUL WEIMER, UNIVERSITY OF COLORADO (GEOLOGY AND BATHYMETRY); LOUISIANA DEPARTMENT OF NATURAL RESOURCES (SHALLOW-WATER WELLS); MMS (DEEP AND ULTRADEEP WELLS, OIL FROM FEDERAL LEASES); ENERGY INFORMATION ADMINISTRATION, OR EIA (U.S. PRODUCTION)

U.S. Gulf oil from federal leases, 1985–2009

Billions of barrels, by depth

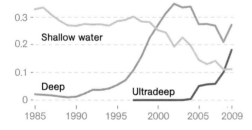

Shallow water
Deep
Ultradeep

0.3
0.2
0.1
0

1985 1990 1995 2000 2005 2009

U.S. domestic oil production, 1985–2009

Billions of barrels

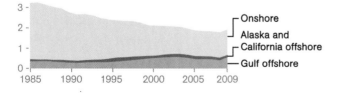

3
2
1
0

1985 1990 1995 2000 2005 2009

Onshore
Alaska and California offshore
Gulf offshore

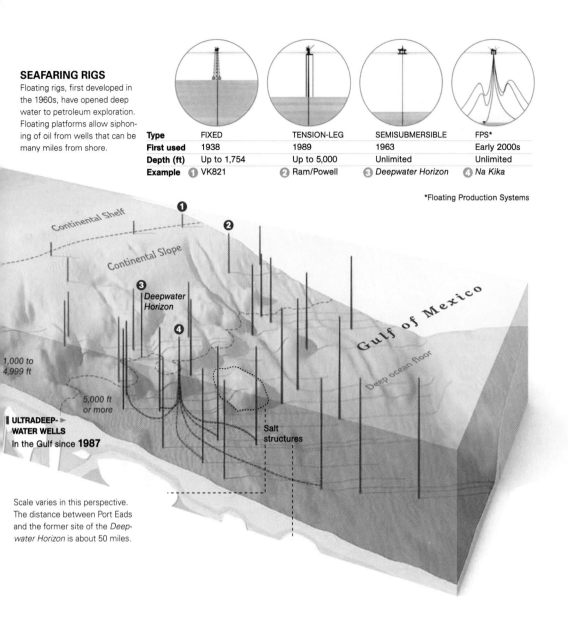

SEAFARING RIGS

Floating rigs, first developed in the 1960s, have opened deep water to petroleum exploration. Floating platforms allow siphoning of oil from wells that can be many miles from shore.

Type	FIXED	TENSION-LEG	SEMISUBMERSIBLE	FPS*
First used	1938	1989	1963	Early 2000s
Depth (ft)	Up to 1,754	Up to 5,000	Unlimited	Unlimited
Example	❶ VK821	❷ Ram/Powell	❸ Deepwater Horizon	❹ Na Kika

*Floating Production Systems

❶

❷ Continental Shelf

Continental Slope

❸ Deepwater Horizon

❹

Gulf of Mexico

Deep ocean floor

1,000 to 4,999 ft

5,000 ft or more

ULTRADEEP-WATER WELLS
In the Gulf since **1987**

Salt structures

Scale varies in this perspective. The distance between Port Eads and the former site of the *Deepwater Horizon* is about 50 miles.

1 NEW DELTA LAND

On the Louisiana coast, new land is being formed in the Atchafalaya River Delta, as river sediment replenishes wetlands. Large-scale diversions of Mississippi and Atchafalaya river waters are proposed to feed the marshes but could interfere with deepwater navigation and key species like oysters.

2 OIL INFRASTRUCTURE

Since the 1940s, oil companies have built thousands of drilling platforms along Louisiana's coast. Tens of thousands of pipelines connect those rigs to shore. The oil industry pumps $70 billion a year into the state; its rigs create a secure habitat for fish. But pipeline canals speed erosion, and the risk of spills is ever present.

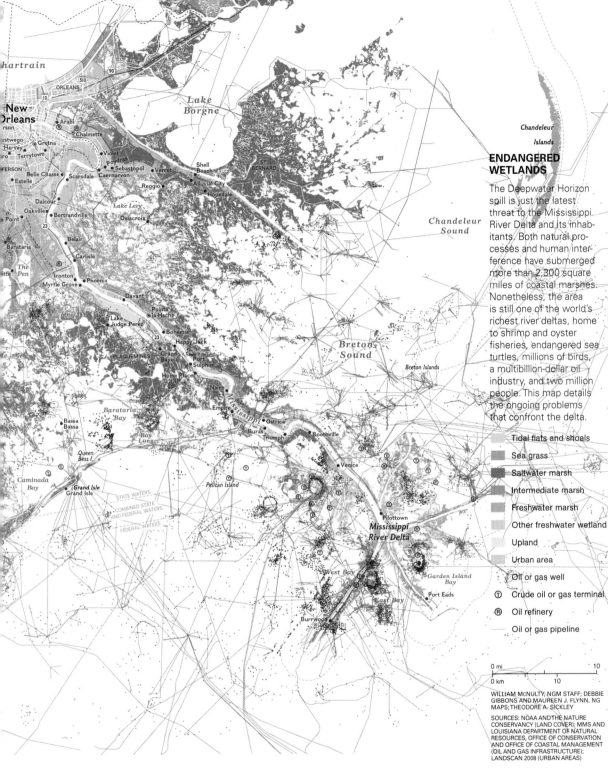

ENDANGERED WETLANDS

The Deepwater Horizon spill is just the latest threat to the Mississippi River Delta and its inhabitants. Both natural processes and human interference have submerged more than 2,300 square miles of coastal marshes. Nonetheless, the area is still one of the world's richest river deltas, home to shrimp and oyster fisheries, endangered sea turtles, millions of birds, a multibillion-dollar oil industry, and two million people. This map details the ongoing problems that confront the delta.

Tidal flats and shoals
Sea grass
Saltwater marsh
Intermediate marsh
Freshwater marsh
Other freshwater wetland
Upland
Urban area
Oil or gas well
Crude oil or gas terminal
Oil refinery
Oil or gas pipeline

0 mi   10
0 km   10

WILLIAM McNULTY, NGM STAFF; DEBBIE GIBBONS AND MAUREEN J. FLYNN, NG MAPS; THEODORE A. SICKLEY

SOURCES: NOAA AND THE NATURE CONSERVANCY (LAND COVER); MMS AND LOUISIANA DEPARTMENT OF NATURAL RESOURCES, OFFICE OF CONSERVATION AND OFFICE OF COASTAL MANAGEMENT (OIL AND GAS INFRASTRUCTURE); LANDSCAN 2008 (URBAN AREAS)

❸ SALTWATER INTRUSION

As wetlands sink and fragment, salt water slips farther inland, killing the freshwater marshes that make up 81 percent of Mississippi River Delta wetlands, home to diverse plants and animals. Canals dug to accommodate oil pipelines and ships speed salt water inland with the tides; faster currents increase erosion.

❹ LIFELESS WATERS

Each summer a "dead zone" of oxygen-starved water develops along the coast. Algae blooms, fed by nitrogen and phosphorus from animal waste and fertilizers from midwestern farms, create this zone, which averages about 6,000 square miles. The dead zone threatens the Gulf's rich coastal fisheries.

❺ VANISHING SHORELINE

Fragile marshland soils need replenishment with sediment and nutrients, but levees built for flood control and navigation shoot those substances out to sea. Draining swamps for development and pumping groundwater cause ground to subside, drowning marsh plants and creating expanses of open water.

AN OILY STAIN

Winds and currents spread surface oil, contaminating more than 625 miles of coastline, most in Louisiana. The spill prompted a fishing ban in one-third of federal waters (partly rescinded in late July) and a massive and ongoing cleanup effort. Experts believe much of the oil never reached the surface and remains in voluminous and elusive underwater plumes.

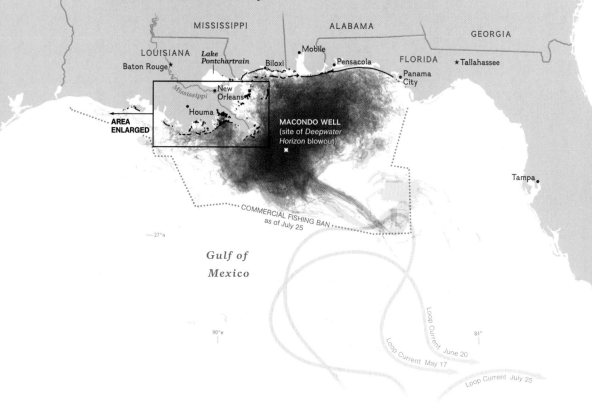

THE BATTERED GULF COAST

Two centuries of efforts to tame the Mississippi River with levees, pumps, and channels have left its vast wetlands ecosystem dwindling and on the verge of collapse. "We know there was a crisis in the Gulf prior to what happened April 20," Tom Strickland, an assistant secretary of the interior, said after the Deepwater Horizon spill. Coastal-restoration plans have been authorized by Congress but are not yet under way. They include breaking open levees to restore the flow of rivers to marshlands. Environmentalists are lobbying to apply oil spill penalty funds to restoration.

(Continued from page 111) shrimps in the Louisiana marshes, but on creatures throughout the region, everything from zooplankton to sperm whales. He's particularly concerned about bluefin tuna, which spawn only in the Gulf and in the Mediterranean; the tuna population was already crashing due to overfishing. "There is a tremendous amount of highly toxic material in the water column, both at the surface and below, moving around in one of the most productive ocean basins in the world," MacDonald said.

During their June cruise Joye's team sampled water within a mile of the *Discoverer Enterprise*, close enough to hear the apocalyptic roar of its (Continued on page 118)

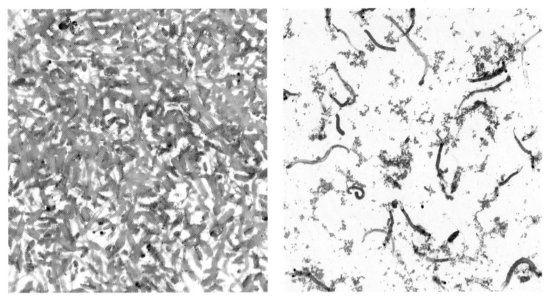

Waters sampled about 35 feet deep on June 28 support a thriving population of tiny crustaceans called copepods (left). Twenty feet farther below was a hypoxic layer almost devoid of life (right). Deep waters are more likely to remain hypoxic.

© David Littschwager/National Geographic Stock

The Spill's Unseen Toll Three formaldehyde-filled jars tell a tale of diminishing life in a water column about 90 miles north of the well. From left to right, the May 4 sample, collected by the Dauphin Island Sea Lab, Ala., shows a normal amount of plankton—minute plants and animals that are the foundation of the ocean's food chain. The June 2 jar holds only 40 percent of the first. The June 28 jar is down to 10 percent. Plankton cannot survive as waters become hypoxic—depleted of oxygen. The probable cause in this case: microbes digesting oil and methane gas from the spill.

© David Littschwager/National Geographic Stock

(Continued from page 116) huge methane flare. Researchers and crew members stood on the back deck of the *Walton Smith* and quietly took pictures. The caustic vapors of oil, diesel, and asphalt burned their lungs. As far as the eye could see, the cobalt blue waters of the deep Gulf were stained brownish red. When Joye went back inside she was in a pensive mood.

"The *Deepwater Horizon* incident is a direct consequence of our global addiction to oil," she said. "Incidents like this are inevitable as we drill in deeper and deeper waters. We're playing a very dangerous game here. If this isn't a call to green power, I don't know what is."

Americans burn nearly 20 million barrels of oil a day. In early August the U.S. Senate adjourned for the summer without taking up an energy bill.

Career Investigation

America has an ever-growing thirst for oil—crude oil that is. Every year the demand for petroleum grows by leaps and bounds. To satisfy our desire to keep our fossil fuel engines purring down the road, we need more petroleum (or an adequate "green" substitute) resources.

Listed here are several possible careers to investigate. You may also be able to find a career path not listed. Choose three possible career paths and investigate what you would need to do to be ready to fill one of those positions.

Among the items to look for:

- Education—What college degree is necessary for this job? What would you major in? Should you have a minor?

- Working conditions—What is the day to day job like? Will you have to be out in the field all day, or is it a desk job? Is the work strenuous? Are you working on a drilling rig at sea, away from your family for 3–6 months at a time? Is the job "hazardous"?

- Pay scale—What is the average "starting pay" for the position you are seeking? Don't be fooled by looking at "average pay" which may include those who have been working for 20+ years.

- Is the job located in the states, or is there a possibility to travel to other parts of the world?

- Are there any other special requirements?

Biochemist
Biologist
Biotechnologist
Botanist
Chemical Engineer
Chemical Engineering Technician
Chemist
Data Analyst/Statistician
Ecologist
Food Scientist/Food Technologist
Geoscientist

Geotechnician
Leakage Operative
Marine Craftsperson
Microbiologist
Oceanographer
Oil Rig Plumber
Oil Rig Worker
Petroleum Engineer
Physicist
Plumber
Project Manager
Research Scientist
Structural Engineer
Technical Surveyor
Zoologist

Team Building Activity

You will be assigned to a team of four or five students. The team will comprise a panel that will examine the incidents at the Deepwater Horizon oil spill and make recommendations to the government and industry leaders on methods to prevent similar incidents.

The team may also take a stance on whether or not Gulf deepwater drilling should be allowed. The team will be expected to justify its decision.

The audience will be expected to ask relevant questions of the panel.

Writing Assignment

After reading the article "The Gulf of Oil: The Deep Dilemma" and conducting your own research, write a brief paper discussing the "need" for deepwater oil drilling in the Gulf of Mexico to handle America's thirst for petroleum products.

In addition to your own ideas and thoughts, please discuss:

- Oil companies are also considering drilling off the Alaskan coast to tap into large oil reserves; do you think this is a good idea? Be sure to explain your rationale.

- Oil companies have also identified large oil reserves (up to 500 million barrels) in the northwest (North Dakota, South Dakota, and Montana), as well as massive deposits in Canada. These remote tracts are on largely unspoiled, undeveloped land. Should drilling be allowed in these areas to satisfy America's thirst for oil? Be sure to explain your rationale.

NEXT: SPACE ELEVATOR

By Luna Shyr

CAN THE BASIC THEORY OF SOUND
ALLOW US COLONIZE MARS
IN THE NEAR FUTURE?

A tethered cargo hold could ferry goods to space.

In 2009 NASA's Andrew Petro watched as a laser-powered robotic device climbed up a cable more than half a mile long above the Mojave Desert. A winner in the agency's Centennial Challenges program—competitions designed to stimulate innovative research—the setup demonstrated the potential of wireless power transmission. That, along with work on super-strong materials, is creating fresh hope for a vision long the realm of science fiction: an elevator that can carry cargo, and possibly people, thousands of miles into outer space.

First described in 1960, the space elevator was also the subject of Arthur C. Clarke's The Fountains of Paradise. Construction is still far from viable, but the basic theory is sound, says Petro. Both power beaming and strong tether materials—crucial aspects of the elevator concept—are featured in NASA's contests and the annual Space Elevator Conference. Another boon was the successful production in 1991 of carbon nanotubes, one of the strongest materials known. But making them suitable for a tether remains a challenge. So why a space elevator at all? Once built, say advocates, it would enable high-volume shipping at a lower cost than rockets. And once that's possible, the next stop could be colonizing Mars.

Adapted from "Next: Space Elevator" by Luna Shyr: National Geographic Magazine, July 2011.

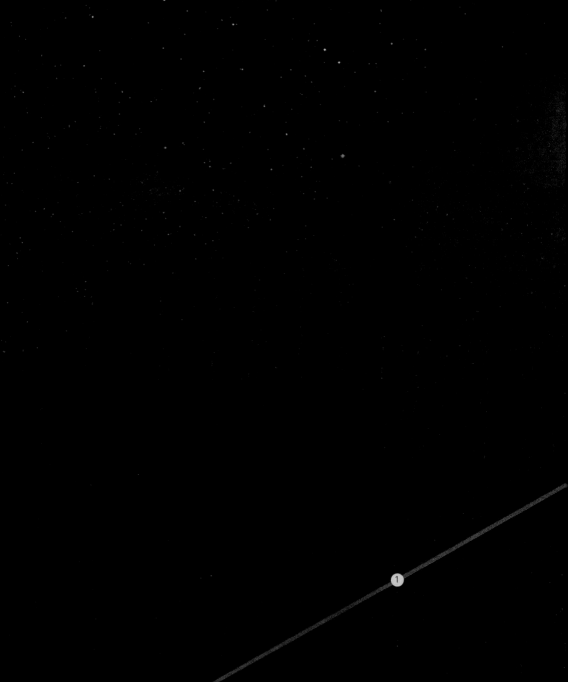

Earth

Climber

Tether

Counterweight

60,000 miles
from Earth

THE BASICS

Like a railcar on a very long track, the climber
(opposite) advances along a ribbon of super-
strong material tethered to a base station on
Earth. As the planet spins, a counterweight
on the far end helps keep the ribbon taut.

ART: STEFAN MORRELL. RESEARCH: ANTHONY SCHICK
SOURCES: BEN SHELEF, PETER SWAN, TED SEMON,
INTERNATIONAL SPACE ELEVATOR CONSORTIUM

Counterweight ▶

Laser light
beamed
from Earth

2

4

3

1

THE TETHER
Carbon nanotubes—
molecular strands of
carbon that are many
times stronger than
steel—show the greatest
promise for a material
strong and light enough
to serve as a tether.
A flat and curved ribbon
shape would help mini-
mize damage from space
debris.

2

**PHOTOVOLTAIC
PANEL**
Photovoltaic cells
on the underside
absorb laser light
beamed from a
base station; the
laser energy is
then converted into
electricity to propel
the climber. Solar
cells on top provide
additional power.

3

THE DRIVE
Like rollers on
a printing press,
multiple wheels
clamp on to the
tether and spin to
move the climber
along. Maintaining
several points of
contact on the
ribbon helps
keep the weight
distributed.

4

CARGO HOLD
Basic models
envision a 14-ton
hold about the size
of an 18-wheeler's
trailer, but the
size could vary,
depending on the
tether. Aluminum,
used in airplanes
and spacecraft, is
a good candidate for
construction.

Career Investigation

What if traveling into space were as simple as stepping into a pressurized elevator compartment and a couple of hours later you would step off onto a space station? Perhaps you are an astronaut going to work for a six-month duty, or maybe you are a tourist going on a space vacation.

Listed here are several possible careers to investigate. You may also be able to find a career path not listed. Choose three possible career paths and investigate what you would need to do to be ready to fill one of those positions.

Among the items to look for:

- Education—Is a college degree necessary for this job? What would you major in? Should you have a minor?

- Working conditions—What is the day to day job like? Will you have to be out in the field all day, or is it a desk job? Is the work strenuous? Are you working on a drilling rig at sea, away from your family for 3–6 months at a time? Is the job "hazardous"?

- Pay scale—What is the average "starting pay" for the position you are seeking? Don't be fooled by looking at "average pay" which may include those who have been working for 20+ years.

- Is the job located in the states, or is there a possibility to travel to other parts of the world?

- Are there any other special requirements?

Biologist
Biomedical Scientist
Biotechnologist
Carbon Nano-tube Engineer
Chemical Engineer
Chemical Engineering Technician
Chemist
Civil Engineer
Civil Engineering Technician
Electrician
Electricity Distribution Worker
Electronics Engineer
Engineering Construction Craftworker
Engineering Construction Technician
Geotechnician
Geophysicist
Laser Engineer
Laser Technician
Physicist
Research Scientist
Welder

Team Building Activity

You will be assigned to a team of four or five students that will host a news conference announcing the opening of a new space elevator that will transport cargo and/or people into space. The operation could be strictly for governmental use or for commercial/for profit business. Your team may use PowerPoint presentations, posters, videos, or other technologies for your presentation.

Writing Assignment

After reading the article "Next: Space Elevator" and conducting your own research, write a brief paper discussing the status of the current space elevator projects and technologies.

In addition to your own ideas and thoughts, please discuss:

- The need for low cost, low energy transport of materials into space.

- The power requirements for a space elevator system and whether or not it can be accomplished using "green" renewable energies.

- The possibilities of reversing the space elevator technologies to bring solar energies back to Earth.

Listed here are a few links to help you in getting started.

Links:

NASA—Space Elevators
http://science.nasa.gov/science-news/science-at-nasa/2000/ast07sep_1/

Spaceward.org
http://www.spaceward.org/elevator

Japan Space Elevator by 2050
http://www.space.com/14656-japanese-space-elevator-2050-proposal.html

Japan Snail Paced Space Elevator—CNET
http://news.cnet.com/8301-17938_105-57383872-1/japan-plans-snail-paced-space-elevator-for-2050/

ANTICIPATION GUIDE

Purpose: To identify what you already know about green building design, to direct and personalize your reading, and to provide a record of what new information you learned.

Before you read "Environment: London's Green Giant," examine each statement below and indicate whether you agree or disagree. Be prepared to discuss your reactions to the statements in groups.

- Green building design is only effective in smaller buildings, such as a house or small apartment building.
- London is well known for its "green" architecture.
- Green architecture can help reduce the amount of required lighting and air conditioning by approximately 50 percent.
- The shape of a building can help the natural ventilation of a building.
- Green architecture is only worthwhile in large cities with tall buildings.
- Some green skyscrapers are being designed to capture rainwater and wastewater for use in the building.

ENVIRONMENT: LONDON'S GREEN GIANT

By Peter Gwin

The "gherkin," one of London's modern buildings.

The Gherkin, Swiss Re or 30 St. Mary Axe
building.
© Richard Nowitz/National Geographic Stock

THE BUILDERS TAKE A
GIANT LEAP INTO THE FUTURE
WHILE SAVING DOLLARS AND MAKING SENSE
FOR THE ENVIRONMENT.

London's Green Giant Nick-named "The Gherkin" for its pickle profile, this distinctive London skyscraper is one of a new crop of buildings designed to reside in better harmony with Mother Nature than previous generations of office towers. The 41-story structure's glass facade and open floor plans allow sun-light to penetrate deep into its interior, reducing the building's reliance on electric lighting. And its curved shape helps direct wind into a natural ventilation system, which mini-mizes the need for air-conditioning. These and other green features may trim the build-ing's total energy consumption to about half that of a comparable conventional structure. Meanwhile, the gherkin's architect, Norman Foster, has turned his attention to New York

A new crop of buildings designed to reside in better harmony with Mother Nature than previous generations of office towers.

City, where his firm is using recycled steel to build Hearst Corporation's new 46-story eco-friendly headquarters, scheduled for completion this summer.

Coming attractions in the green-building movement:

Bank of America Tower Now under construction, the Man-hattan skyscraper will capture and reuse rain and wastewater.

Beijing Olympics Organizers say all new venues for the 2008 Summer Games will incorporate green design.

Guadalajara stadium The Mexican city's new soccer arena will be built within a hill. Outside, its grass-covered slopes will serve as a public park.

Adapted from "Environment: London's Green Giant" by Peter Gwin: National Geographic Magazine, March 2006.

© Richard Nowitz/National Geographic Stock

Career Investigation

More and more buildings are being designed with sustainability in mind. Building owners (whether single family or high-rise urban buildings) look for building designs that will lessen dependency on electrical lighting, energy hungry air conditioning and heating, and even water consumption. In addition to architects, there are numerous careers applicable to "green" building designs.

Listed here are several possible careers to investigate. You may also be able to find a career path not listed. Choose three possible career paths and investigate what you would need to do to be ready to fill one of those positions.

Among the items to look for:

- Education—Is a college degree necessary for this job? What would you major in? Should you have a minor?

- Working conditions—What is the day to day job like? Will you have to be out in the field all day, or is it a desk job? Is the work strenuous? Are you working on a drilling rig at sea, away from your family for 3–6 months at a time? Is the job "hazardous"?

- Pay scale—What is the average "starting pay" for the position you are seeking? Don't be fooled by looking at "average pay" which may include those who have been working for 20+ years.

- Is the job located in the states, or is there a possibility to travel to other parts of the world?

- Are there any other special requirements?

Biofuel Technology Development Manager
Biologist
Cartographer
Chemical Engineer
Civil Engineer
Civil Engineering Technician
Climate Change Analyst
Deep Sea Diver
Emergency Management Specialist
Environmental Compliance Inspector
Environmental Engineer
Environmental Engineering Technician
Environmental Scientist
Fuel Cell Engineer
Hydroelectric Plant Technician
Industrial Engineer
Geographer
Geoscientist
Hydrologist
Marine Biologist
Marine Engineer

Mechanical Engineer
Mechanical Engineering Technician
Methane Gas Generation System
  Technician
Nuclear Engineer
Park Ranger
Petroleum Engineer
Photonics Engineer
Photonics Technician
Solar Energy System Engineer
Solar Energy System Technician
Soil Scientist
Solid Waste Engineer
Sustainability Specialist
Water & Liquid Waste Treatment Engineer
Wind Energy Engineer
Wind Turbine Service Technician

Team Building Activity

You will be assigned to a team of four or five students and asked to think about your town or city's current "green" practices. Prepare a report on what you think is the most green element in your town. Be sure to include what makes this element "green" and where it could be improved to help the environment even more. Present your findings to the other teams. Do you all agree or are there different opinions? Discuss the various elements that each team thought were important.

Writing Assignment

After reading the article "Environment: London's Green Giant" and conducting your own research, write a brief persuasive paper addressed to the building superintendent in the high rise where you live to convince him/her to retrofit your aging building to utilize "green" design features.

In addition to your own ideas and thoughts, please discuss:

- The benefits to the building and residents.

- The benefits to the local ecology.

- Possible design considerations.

- Anticipated cost of the project.

UP ON THE ROOF

By Verlyn Klinkenborg

Photographs by Diane Cook
and Len Jenshel

Purpose: To identify what you already know about urban living and green spaces, to direct and personalize your reading, and to provide a record of what new information you learned.

Before you read about urban rooftop green spaces in "Up on the Roof," examine each statement below and indicate whether you agree or disagree. Be prepared to discuss your reactions to the statements in groups.

- Putting sod or a garden on top of buildings is a relatively new concept.
- A rooftop green space can be an ecological benefit to the inner city.
- Green spaces on buildings are only for businesses.
- Having a sod or garden rooftop means you do not need a regular roof.
- Having a sod or garden rooftop will help provide insulation from the cold of winter and the heat of summer.
- In some places having a garden or sod rooftop is not only a good idea, it is required by law.
- Rooftop green spaces reduce the amount of water being flushed into city drains and sewer systems.
- City restaurants can use a rooftop garden to grow fruits, herbs, and vegetables to be used in the restaurant.
- Rooftop green spaces are not only for urban buildings but are equally effective in suburban and rural areas.
- In addition to being used on homes and businesses, rooftop green spaces can be used on smaller structures like sheds and bus stops.
- Installing a rooftop green space is as easy as putting some soil on the roof and planting grass or plants.
- Having a rooftop green space allows for a habitat to reintroduce itself to an urban setting.

A garland of nature crowns Chicago's City Hall, softening the hard edges of a town famous for steel and stone—and lowering summer temperatures on the roof. Inspired by a worldwide movement, Mayor Richard Daley has made Chicago North America's leading "green roofs" city.

A Manhattan apartment building poses as a country cottage—a retreat former condo developer David Puchkoff built for his family. A verdant roof absorbs rain, reducing runoff, so it's also an environmental gift to New York City, where flooded sewers foul the Hudson River after downpours.

A LOFTY IDEA IS

BLOSSOMING IN CITIES AROUND THE WORLD,

WHERE ACRES OF POTENTIAL GREEN SPACE LIE OVERHEAD.

If buildings sprang up suddenly out of the ground like mushrooms, their rooftops would be covered with a layer of soil and plants.

That's not how humans build, of course. Instead we scrape away the earth, erect the structure itself, and cap it with a rainproof, presumably forgettable, roof. It's tempting to say that the roofscape of every city on this planet is a man-made desert, except that a desert is a living habitat. The truth is harsher. The urban roofscape is a little like hell—a lifeless place of bituminous surfaces, violent temperature contrasts, bitter winds, and an antipathy to water.

But step out through a hatch onto the roof of the Vancouver Public Library at Library Square—nine stories above downtown—and you'll find yourself in a prairie, not an asphalt wasteland. Sinuous bands of fescues stream across the roof, planted not in flats or containers but into a special mix of soil on the roof. It's a grassland in the sky. At ground level, this 20,000-square-foot garden—created in 1995 by landscape architect Cornelia H. Oberlander—would be striking enough. High above Vancouver, the effect is almost disorienting. When we go to the rooftops in cities, it's usually to

look out at the view. On top of the library, however, I can't help feeling that I'm standing on the view—this unexpected thicket of green, blue, and brown grasses in the midst of so much glass and steel and concrete.

Living roofs aren't new. They were common among sod houses on the American prairie, and roofs of turf can still be found on log houses and sheds in northern Europe. But in recent decades, architects, builders, and city planners all across the planet have begun turning to green roofs not for their beauty—almost an afterthought—but for their practicality, their ability to mitigate the environmental extremes common on conventional roofs.

Across town from the library, the Vancouver Convention Centre is getting a new living roof. Just across the street there is a chef's garden on the roof of the Fairmont Waterfront hotel. Across town in another direction, green roofs will go up on an Olympic village

Adapted from "Up on the Roof" by Verlyn Klinkenborg: National Geographic Magazine, June 2011.

INSIDE A LIVING ROOF

A green roof on a commercial building is typically composed of these essential layers.

Vegetation
Water-storing plants such as sedums drink in rain that would otherwise run off a traditional flat roof.

Growing medium
Natural soil weighs too much when waterlogged, so green-roof architects use a soil composite.

Drainage
Excess rainwater filters into a layer of storage cups or pebbles before overflowing into a drain.

During dry periods, this stored water is drawn back up to the roots.

Support
A root barrier and waterproof membrane separate the living-roof system from the insulated building below.

DEPTH RANGES 3-15 INCHES

FILTER FABRIC

STORAGE CUPS

ROOT BARRIER MAT

WATERPROOF MEMBRANE

INSULATION

STRUCTURAL DECK

MARIEL FURLONG, NG STAFF
ART BY DON FOLEY
SOURCE: BARBARA DEUTSCH

Walking paths and conical skylights share the grassy roof of the Art and Exhibition Hall in Bonn, Germany. With government aid, Germans led the development of modern green roofs in the 1960s. Today many cities offer incentives for their use.

being built for the 2010 Winter Olympics. To stand on a green roof in Vancouver—or Chicago or Stuttgart or Singapore or Tokyo—is to glimpse how different the roofscapes of our cities might look and to wonder, Why haven't we always built this way?

Technology is only partly the reason. Waterproof membranes now make it easier to design green-roof systems that capture water for irrigation, allow drainage, support the growing medium, and resist the invasion of roots. In some places, such as Portland, Oregon, builders are encouraged to use living roofs by fee reductions and other incentives. In others—such as Germany, Switzerland, and Austria—living roofs are required by law on roofs of suitable pitch.

And, increasingly, researchers such as Maureen Connelly—who runs a green-roof lab at the British Columbia Institute of Technology—are studying the practical benefits green roofs offer, helping quantify how they perform and providing an accurate measure of their ability to reduce storm-water runoff, increase energy efficiency, and enhance the urban soundscape. There is beginning to be a critical mass of green roofs around the world, each one an experiment in itself.

Another factor driving the spread of green roofs is our changing *(Continued on page 144)*

Hardy sedums, insects, and birds rule the roof at a hospital in Basel, Switzerland, where foliage is mandatory on new flat roofs. "If we steal the ground for a building," says International Green Roof Association director Wolfgang Ansel, "we can give it back to nature on the roof."

A bus shelter in downtown San Francisco (right) and a shed at the Oregon Garden in Silverton, Oregon (above), support tiny living roofs. They're intended to plant a seed in the minds of Americans. Diane Loviglio, who planned the bus shelter roof, hopes ordinary passersby will see "a viable home-improvement idea." The sustainable-design activist wanted to showcase the green-roof idea, less familiar in the U.S. than in much of Europe, "at street level," she says, "so people don't have to tour a giant industrial building to understand it."

(Continued from page 141) idea of the city. It's no longer wise or practical or, for that matter, ethical, to think of the city as the antithesis of nature. Finding ways to naturalize cities—even as nature itself becomes more urbanized—will make them more livable, and not only for humans.

Living roofs remind us what a moderating force natural biological systems are. During the summer, daytime temperatures on conventional asphalt rooftops can be almost unbelievably high, peaking above 150°F and contributing to the overall urban heat-island effect—the tendency of cities to be warmer than the surrounding region. On green roofs the soil mixture and vegetation act as insulation, and temperatures fluctuate only

mildly—hardly more than they would in a park or garden—reducing heating and cooling costs in the buildings below them by as much as 20 percent.

When rain falls on a conventional roof, it sheets off the city's artificial cliffs and floods down its artificial canyons into storm drains—unabsorbed, unfiltered, and nearly undeterred. A living roof works the way a meadow does, absorbing water, filtering it, slowing it down, even storing some of it for later use. That ultimately helps reduce the threat of sewer overflows, extends the life of a city's drain system, and returns cleaner water to the surrounding watershed. London, for example, is already planning (Continued on page 148)

The windows of these traditional London row houses once opened onto a decrepit sausage factory. Now residents face a wildflower meadow blooming on top of architect Justin Bere's new home. Insulation provided by the green roof helps make Bere's solar-powered house energy efficient.

(Continued from page 145) for a future that may well see more street flooding, and the city is considering how living roofs could moderate the threat.

Above all, living roofs are habitable. They recapture what is now essentially negative space within the city and turn it into a chain of rooftop islands that connect with the countryside at large. Species large and small—ants, spiders, beetles, lapwings, plovers, crows—have taken up occupancy on living roofs. The list includes Britain's black redstarts, a bird that colonizes the rubble of abandoned industrial sites, a habitat being lost to redevelopment. The solution fostered by Dusty Gedge, a British wildlife consultant and a driving force behind green roofs in the United Kingdom, is to create living rooftop habitat out of the same rubble. And it's not just a matter of making new or replacing existing habitat. In Zürich, Switzerland, the 95-year-old living roof of a water-filtration system serves as a refuge for nine species of native orchids eradicated from the surrounding countryside when their meadow habitat was converted to cropland.

Proponents of living roofs argue that they have met most, if not all, of the technical challenges involved in grafting a biological layer onto the top of buildings of almost any scale: everything from a vegetable stand or bus stop to the ten-acre roof of Ford's truck plant in Dearborn, Michigan. While the average cost of installing a green roof can run two or three times more than a conventional roof, it's likely to be cheaper in the long run, thanks largely to energy savings. Vegetation also shields the roof from ultraviolet radiation, extending its life. And it requires a different kind of care, akin to low-maintenance gardening.

There are still philosophical challenges to be met, many of them having to do with the very idea of what a roof should be and how it should perform. Clients tend to want roofs that are easy to maintain and are uniformly green year-round, perpetual lawns in the sky, not seasonal

There is beginning to be a critical mass of green roofs around the world, each one an experiment in itself.

grasslands. Builders and architects tend to want interchangeable, standardized, universal solutions, the kind of green-roof systems now being offered by some of the big corporate players in the living-roofs industry.

A living roof, though, is not just a biological alternative to a dead roof. It requires a different way of thinking altogether. A standardized green roof such as a carpet of sedums is better than a conventional roof, but it's possible to build living roofs that are even more environmentally beneficial—locally grown, so to speak. The goal for some researchers now is to find ways to build living roofs that are ecologically and socially sound in every respect: low in environmental costs and available to as many people as possible.

Stephan Brenneisen, a Swiss scientist and a strong advocate for the biodiversity potential of living roofs, says simply, "I have to find easy, cheap solutions using materials that come from the region." That means less reliance on plastics and other energy-intensive materials between the roof structure and the plants themselves. What matters isn't only whether living roofs work. It's how to make them work in the most sustainable way, using the least energy while creating the greatest benefit for the human and nonhuman habitat.

Last fall, I climbed onto the roof of the 15-story Portland Building in downtown Portland, Oregon. My guide was Tom Liptan, the city's Ecoroof Program Manager and a self-confessed storm-water nerd, who began his experiments with green roofs by building one on his own garage in 1996. We walked to the parapet across plantings of sedums and fescues and looked down at the roof of Portland's city hall several stories below us. It has a conventional black tar roof, the kind of roof we have taken for granted for decades. But as part of Portland's Grey to Green project—a plan for sustainable storm-water *(Continued on page 152)*

Wasted space in the modern metropolis may become productive "farmland" thanks to advances in waterproofing green roofs. Some of the rice used to brew Japan's popular Hakutsuru sake grows atop the company's Tokyo office. A chef at Vancouver's Fairmont Waterfront hotel harvests apples ripening among skyscrapers (right). Hotel accountants say the roof garden produces fruits, vegetables, herbs, and honey worth about $16,000 annually.

It looks like a prehistoric ruin overgrown with ferns, but it may be the next step in foliage-based architecture. The Vancouver Aquarium's Living Wall varies the classic vine-covered trellis with movable vertical planter boxes and built-in irrigation, a design adaptable to many settings.

In the heartland of American industry—Dearborn, Michigan, where Henry Ford revolutionized manufacturing—nature makes a comeback on one of the world's largest green factory roofs (above). Ford Motor Company installed sedums on the 10.4-acre expanse to reduce runoff from the site. New York's Empire State Building gleams in the windows of architectural firm Cook + Fox (right). Specialists in green buildings, the designers wanted their own space to reflect the fact that more plants in more places make for more livable cities.

(Continued from page 148) management—that building will soon be retrofitted with a living roof. "The employees want it," Liptan said.

In the history of that municipal building, how often had the people who worked there ever thought about that black tar roof looming over their heads? Once the living roof is completed, they may visit it only rarely, but they won't forget that it's there, adding habitat to the city center, filtering the rain, moderating temperatures.

It reminded me of something Stephan Brenneisen said: "People feel happier in a building where we've given something back to nature."

Think of the millions of acres of unnatural rooftops around the globe. And now imagine returning some of that enormous human footprint to nature—creating green spaces where there was once only asphalt and gravel. If a certain sum of human happiness is the by-product, who's to complain?

Career Investigation

Imagine flying high over a metropolitan city. What would you expect to see when you look down? Would it be barren black tar rooftops devoid of life? Wouldn't it be nice to see green living rooftops hosting a variety of plants that are helping the ecology?

Listed here are several possible careers to investigate. You may also be able to find a career path not listed. Choose three possible career paths and investigate what you would need to do to be ready to fill one of those positions.

Among the items to look for:

- Education—Is a college degree necessary for this job? What would you major in? Should you have a minor?

- Working conditions—What is the day to day job like? Will you have to be out in the field all day, or is it a desk job? Is the work strenuous? Are you working on a drilling rig at sea, away from your family for 3–6 months at a time? Is the job "hazardous"?

- Pay scale—What is the average "starting" pay for the position you are seeking. Don't be fooled by looking at "average pay" which may include those who have been working for 20+ years.

- Is the job located in the states, or is there a possibility to travel to other parts of the world?

- Are there any other special requirements?

Architect
Architectural Technician or Technologist
Biologist
Biomedical Scientist
Biotechnologist
Botanist
Bricklayer
Building Technician
Civil Engineer
Civil Engineering Technician
Construction Manager
Ecologist
Facilities Manager
Landscape Manager
Landscape Scientist
Landscaper

Mechanical Engineering Technician
Microbiologist
Plumber
Roofer
Sheet Metal Worker
Structural Engineer
Welder

Team Building Activity

Your team will design and build a scale model living roof system. The design of the roof system must fit in with a chosen urban-metropolitan locale. As you design your living rooftop, be sure to include all of the design considerations (including drainage and support) which you have learned in reading the article "Up on the Roof" as well as your own research.

Writing Assignment

After reading the article "Up on the Roof" and conducting your own research, write a brief persuasive paper addressed to the building superintendent in the high rise where you live to convince him/her to retrofit your "dead" rooftop into a viable living rooftop.

In addition to your own ideas and thoughts, please discuss:

- The benefits to the building and residents.

- The benefits to the local ecology.

- Possible design considerations.

- Anticipated cost of the project.

Listed here are a few links to help you in getting started.

Links:

Green Rooftops in Beijing
http://www.china.org.cn/
environment/2012-04/02/
content_25052715.htm

Green Roof Presentation
http://community.seas.columbia.edu/cslp/
reports/summer07/greenroofGreen_roof_
final_pdr.pdf

Green Roof Technology
http://extension.ucdavis.edu/unit/green_
building_and_sustainability/pdf/resources/
green_roof.pdf

Green Roof Volunteer Guide
http://www.volunteerguide.org/hours/
service-projects/green-roof

Green Roofs
http://www.greenroofs.org/index.php/
about-green-roofs/2577-aboutgrnroofs

Green Roof Benefits
http://curiosity.discovery.com/question/
green-roofs-good-for-environment